# STATISTICS FOR EARTH AND ENVIRONMENTAL SCIENTISTS

# Statistics for Earth and Environmental Scientists

*Editor*

**Jacek Fabian**

scitus
academics

**Statistics for Earth and Environmental Scientists**
Edited by **Jacek Fabian**

ISBN: 978-1-68117-571-3
Library of Congress Control Number: 2016943006


SCITUS Academics LLC,
www.scitusacademics.com
Box No. 4766, 616 Corporate Way,
Suite 2, Valley Cottage,
NY 10989



**Notice**



Printed in United States of America on Acid Free Paper

# PREFACE

Earth scientists are often involved in taking observations on the earth's surface and its interior. Earth scientists are confronted with many challenges. Phenomena of interest are often deep within the earth surface, in oceans, or in the atmosphere. Collecting data often is costly, time consuming, and hampered by limited access. Geologic events occur at irregular intervals, often widely spaced. Decisions made on the basis of studies in the earth sciences can have significant long term consequences. An understanding of data used in decision making is critical. Exploratory data analysis provides insight into data quality, variability, and structure. Constructing models provides a basis for understanding complex geologic processes and for making inferences. Since earth scientists depend largely on observations, particularly on observations where there is a large portion of uncertainty, Statistics plays a major role in drawing inferences concerning the earth and its interior.

*Statistics for Earth and Environmental Scientists* provides comprehensive theoretical and practical knowledge of statistical applications for solving real-world environmental problems. This book is envisioned for students and practitioners of the earth and environmental sciences who want to practise statistical tools to unravel real problems. It offers a variety of tools that are used across earth science disciplines. Statistical methods need to be understood because today's interesting problems are complex and include uncertainty. These complex problems take account of energy resources, climate change, and geologic hazards.

# TABLE OF CONTENTS

7

# OPTIMIZING GROUNDWATER MONITORING NETWORKS USING INTEGRATED STATISTICAL AND GEOSTATISTICAL APPROACHES

**Jay Krishna Thakur**

Environment and Information Technology Centre—UIZ,
Neue Grünstraße 38, Berlin 10179, Germany;

## ABSTRACT

The aim of this work is to investigate new approaches using methods based on statistics and geo-statistics for spatio-temporal optimization of groundwater monitoring networks. The formulated and integrated methods were tested with the groundwater quality data set of Bitterfeld/Wolfen, Germany. Spatially, the monitoring network was optimized using geo-statistical methods. Temporal optimization of the monitoring network was carried out using Sen's method (1968). For geostatistical network optimization, a geostatistical spatio-temporal algorithm was used to identify redundant wells in 2- and 2.5-D Quaternary and Tertiary aquifers. Influences of interpolation block width, dimension, contaminant association, groundwater flow direction and aquifer homogeneity on statistical and geostatistical methods for monitoring network optimization were analysed. The integrated approach shows 37% and 28% redundancies in the monitoring network in Quaternary aquifer and Tertiary aquifer respectively. The geostatistical method also recommends 41 and 22 new monitoring wells in the Quaternary and Tertiary aquifers respectively. In temporal optimization, an overall optimized sampling interval was recommended in terms of lower quartile (238 days), median quartile (317 days) and upper quartile (401 days) in the research area of Bitterfeld/Wolfen. Demonstrated methods for improving groundwater monitoring network can be used in real monitoring network optimization with due consideration given to influencing factors.

## KEYWORDS

groundwater monitoring network optimization; factors influencing monitoring network optimization; quaternary and tertiary aquifers

## 1. INTRODUCTION

In regions with frequent water stress and large aquifer systems, groundwater is often used as an additional water resource. If extensive groundwater abstraction exceeds natural groundwater recharge for longer times, overexploitation or persistent groundwater depletion occurs [1].

Population growth and groundwater depletion are two of the most significant dangers to global water stability. It is evident from the trending water scarcity and continual groundwater pollution throughout the world that the existing policies fail to protect this vital resource [2,3].

In almost all major water resource management programs, groundwater monitoring is required [4]. Ground water monitoring involves an array of sequential processes ranging from long term standardized measurement, observation, status and trend evaluation to the final reporting of processed data in order to meet the objectives of monitoring programmes [5,6].

Monitoring networks with sets of strategically located groundwater monitoring wells collecting prerequisite data [7] that, in turn, provide insights for strategic planning and decision making [8]. Since complex infrastructure is crucial for entire data analysis and processing activities, ground water monitoring acts as a significant burden at many ground water contamination sites [9,10].

The last three decades has achieved great improvement in groundwater monitoring strategies, the majority of which have strongly focused on statistical approach for monitoring locations and sample size [11,12,13]. However, relatively little work is in place with optimization considering multiple contaminants in multiple aquifers.

For proper design of monitoring strategy as per management objectives, it is necessary that required information be extracted from measured variables [4,14]. Thereafter, the monitoring strategy can be derived based on optimization of the groundwater monitoring network and concurrent considerations of the uncertainty associated with spatial and temporal variability [15,16].

In the past, various studies have specified various methods for monitoring network optimization [17,18,19,20,21,22] but have failed to incorporate the primary factors that influence the methods for so. This study attempts to thoroughly analyze the influence of various factors on the optimization method, which can have a remarkable role in the decision about redundancy and necessity of new wells in the monitoring network.

The aim of this research is to identify and investigate new approaches using statistical and geostatistical methods for groundwater monitoring network optimization in order to improve groundwater monitoring strategies. The factors influencing groundwater monitoring optimization methods will be analyzed and new and improved methods will be integrated by testing those with groundwater monitoring network optimization at contaminated sites.

# 2. STUDY AREA AND DATA USED

## 2.1. Study Area

In order to address the research questions and test the developed approaches and methods, Bitterfeld/Wolfen, located in the Federal State Saxony-Anhalt, Germany, was selected as a study area (Figure 1). Yet again, to make precise application of improved and developed methods in this study area, an area of about 100 $km^2$ in urbanized zone of Bitterfeld/Wolfen (latitude 51°35′30″–51°41′30″ and longitude 12°14′10″–12°20′0.5″) consisting of a long term monitoring network (LTM) of 357 wells in the Tertiary and 462 wells in the Quaternary aquifer was selected.

In terms of geography, the western part of the study area is covered by glacial outwash sediments, whereas the flood plain of the Mulde river constitutes the eastern part of research area [23]. The geological setting of the research area reveals the presence of pre-tertiary rocks (overlaid by Cenozoic sediments), separated hydro-geologically from one another by clay layer at the depth of 50–70 m [24]. An upper Quarternary aquifer system, lower Tertiary aquifer system and pre-tertiary basement constitutes a geological cross section in the research area as portrayed by the Bitterfeld/Wolfen model Stollberg, Gossel *et al.* [25].

## 2.2. Nature and Availability of Data

An available data set of physicochemical parameters (α-Hexachlorocyclohexane (α-HCH), Monochlorobenzene (MCB), temperature, pH-value, oxidation potential (Eh), sulphate ($SO_4^{2-}$), sulphite ($SO_3^{2-}$), nitrate ($NO_3^-$), ammonium ($NH_4^+$) and iron ($Fe^{3+}$)) from the period 2003 to 2009 were used for LTM network optimization. Assessory information like name of the groundwater observation well, coordinates, elevation data of the monitoring well, screen depth, stratigraphic geological layer, stratigraphic horizon (Quaternary (Q), Tertiary (T) and Quaternary-Tertiary (Q-T)), the date and time of sampling and the name of the analysing laboratory (Table 1) were also available. In the table, the 9th column gives the total number of wells and samples during the overall period 2003–2009.

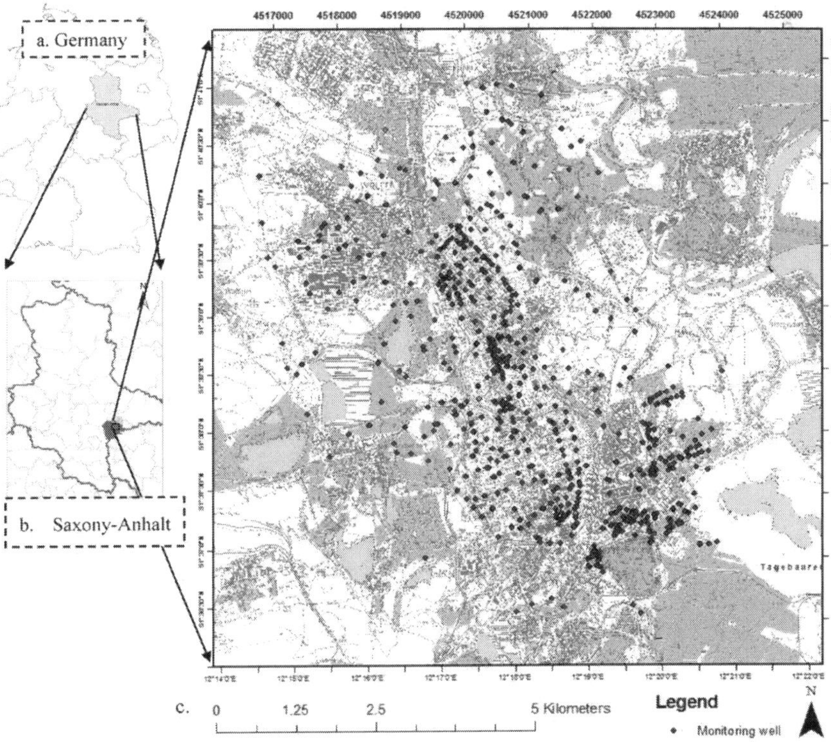


**Figure 1.** Location map of the study area in Bitterfeld, Federal State of Saxony-Anhalt, Germany (**a**) Saxony-Anhalt in eastern Germany, (**b**) Federal State of Saxony-Anhalt, (**c**) study area of 100 km$^2$ used for monitoring network optimization in Bitterfeld/Wolfen showing location of groundwater monitoring wells).

## 3. METHODS

Design and improvement of monitoring strategies requires the description of its components. The components of groundwater monitoring strategy, analyzed for monitoring scenario in the study area, is represented in the logic diagram (Figure 2).

**Table 1.** Number of wells and samples from 30 September 2003 to 15 December 2009, showing number of wells monitored each year for physicochemical parameters.

| Year | 2003 | 2004 | 2005 | 2006 | 2007 | 2008 | 2009 | 2003–2009 |
|---|---|---|---|---|---|---|---|---|
| No of well | 477 | 579 | 496 | 663 | 682 | 521 | 38 | 827 |
| No of sample | 796 | 749 | 729 | 847 | 711 | 519 | 38 | 4389 |
| pH | 786 | 719 | 720 | 865 | 501 | 519 | 38 | 4148 |
| Eh | 787 | 701 | 703 | 841 | 501 | 519 | 38 | 4090 |
| $NO_3^-$ | 787 | 709 | 719 | 866 | 414 | 519 | 31 | 4045 |
| $SO_3^{2-}$ | 796 | 703 | 700 | 864 | 424 | 519 | 31 | 4037 |
| $SO_4^{2-}$ | 795 | 729 | 720 | 866 | 424 | 519 | 31 | 4084 |
| $NH_4^+$ | 717 | 729 | 720 | 866 | 420 | 519 | 31 | 4002 |
| $Fe^{2+}$ | 0 | 8 | 102 | 18 | 0 | 0 | 0 | 128 |
| $Fe^{3+}$ | 772 | 709 | 694 | 841 | 475 | 519 | 38 | 4048 |
| α-HCH | 772 | 682 | 698 | 823 | 500 | 519 | 38 | 4032 |
| MCB | 735 | 678 | 739 | 2156 | 1954 | 2307 | 2097 | 10,666 |

Among the groundwater physicochemical properties and contaminants concentration data, representative contaminants were first identified and then prioritized for inclusion in the monitoring strategy. Their associated uncertainties along with spatial and temporal variance were incorporated.

## 3.1. Selecting the Representative Variable

Identification of monitoring components in the research area was done through Principal Component Analysis (PCA). PCA of the normalized variables was performed to extract significant PCs. The variables were standardized through z-scale transformation to avoid misclassification due to difference in data dimensionality [26,27]. PCA provides information about the principle variables responsible for spatial variation in groundwater quality [28].

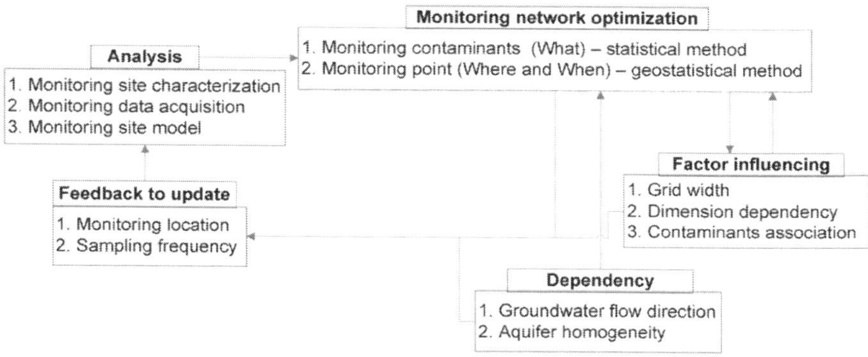


**Figure 2.** Logic diagram showing components of groundwater monitoring strategies as a circular continuous process.

Measures of descriptive statistics in conjunction with statistical modelling of all available water quality and contaminant concentration data were used to identify the representative contaminants. World Health Organization (WHO) standards for drinking water and maximum contamination limit (MCL) of pollutants and contaminant concentration as well as associated human health risk were considered in the course of identification of the representative contaminants [29,30].

## 3.2. LTM Network Optimization

Necessary information about contaminants, including their properties, hydro-geochemical properties of aquifers, and other related information, are the pre-requisites for optimization of LTM network. Representative contaminants in the study area were MCB, $SO_4^{2-}$ and α-HCH, which was identified based on PCA that was used for monitoring network optimization. MCB and $SO_4^{2-}$ represent organic and inorganic contamination respectively in the research area [31,32]. α-HCH, which is an organochloride and one of the isomers of hexachlorocyclohexane (HCH), represents pesticides [33,34].

### 3.2.1. Spatial Optimization Methods

Redundant wells were predicted using Geostatistical Temporal-Spatial algorithm (GTS) on a concrete basis that the remaining wells represent the same amount of statistical information about the underlying plume [35,36]. GTS concept considers a well or its group redundant if it plays insignificant contribution to the interpolated map of contaminant plumes. The investigation steps involved in locating essential and redundant wells are shown in Figure 3.

An interpolated contaminant concentration plume was generated with input data of MCB, α-HCH and $SO_4^{2-}$ at specified locations of the well. The maximum concentration limit (MCL) for each contaminant *i.e.*, 100 μg/L, 0.2 μg/L and 0.25 g/L respectively for MCB, α-HCH and $SO_4^{2-}$ [30] was given as input indicator limit to the statistical algorithm. The contaminant concentration was converted into an indicator value as

$$i_C(t) = 1 \text{ if } z(t) > C \text{ and } 0 \text{ if } z(t) \leq C \tag{1}$$

where, $i_C(t)$ is the indictor value, z(t) is the concentration at time t, and C is the maximum contamination limit (MCL).

Once again, to take into account the correlation based on distance and direction between pairs of sampling locations for each contaminant, the empirical variogram was modelled. The empirical spatial variogram, γ, as a function of lag distance can be expressed as:

$$\gamma(\Delta h) = \frac{1}{2N(\Delta h)} \sum_{i,j \left\| x_j - x_i \right| - \Delta h \,|\leq \varepsilon} \left[ i_C(x_j) - i_C(x_i) \right]^2 \tag{2}$$

where $\Delta h$ is the targeted lag distance and may depend on direction; $x_i$ and $x_j$ are the $i$th and $j$th monitoring well locations; and $N(\Delta h)$ is the number of indicator pairs contributing to the summation for lag $\Delta h$.

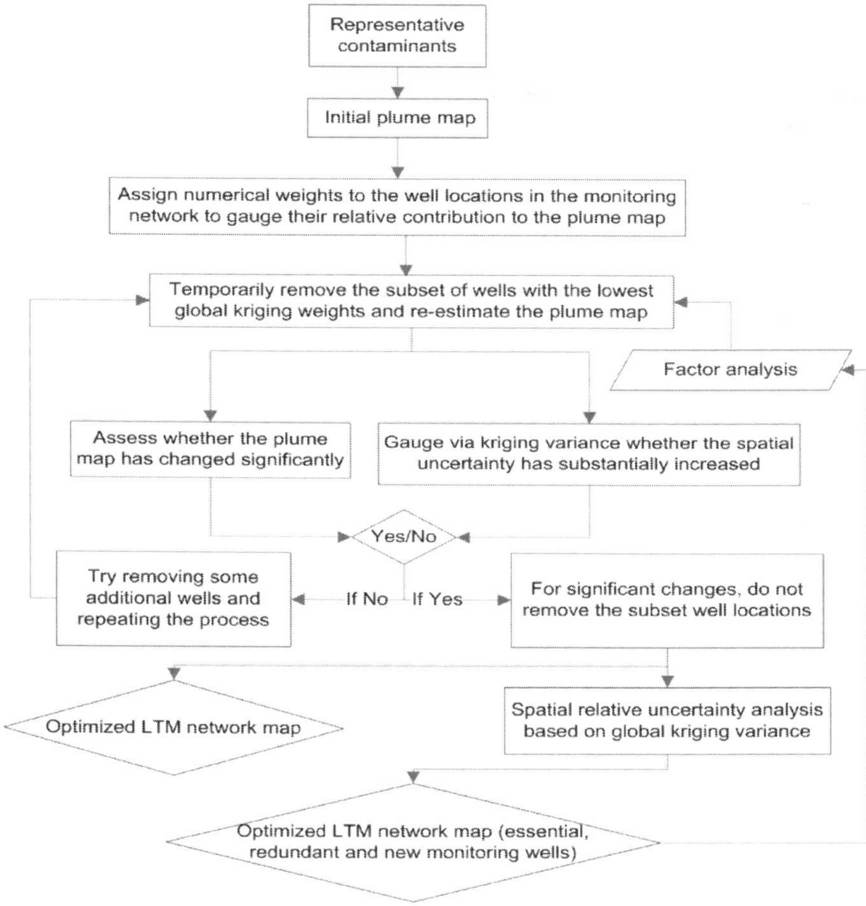


**Figure 3.** Research steps to locate essential and redundant wells and to propose new wells in the existing monitoring network (Modified from Cameron and Philip M. Hunter (2010) [36]).

Numerical weight was given to the wells based on global kriging and a plume map was generated for the monitoring network. For this process, two intermediate computations were carried out:

(i) the local kriging weights of sampled locations which were processed to generate a "global" interpolation weight for each well [37]; and (ii) the local kriging estimation variance which indicate the relative uncertainty of the local block estimates in comparison to the estimates at other blocks [36].

Spatial-temporal algorithm incorporates series of procedural steps to estimate global interpolation weight and the local kriging estimation variance. At first, the study area is divided into sets of non-overlapping blocks. The set of closest sampled locations is then located at each block with the help of simple search algorithm. Then, local kriging weights ($\lambda^B$) is computed for each well based on the spatial configuration of known indicator value within and surrounding the block as well as spatial correlation between average block location and each known indicator by using modelled spatial covariance function.

A block indicator estimate is then generated with the combination of local weights and indicator values. A block indicator estimate consist of weighted average of the $n(x_B)$ indicators located within the search radius of the block ($x_B$), which is estimated as [36]:

$$i_C(x_B) = \sum_{i=1}^{n(x_B)} \lambda_i^B i_C(x_i) \tag{3}$$

for $x_i \in$ search radius of block $x_B$.

Global interpolation weights ($\lambda^G$) is generated by averaging the local kriging weights to the given well that can be used to estimate the well's overall contribution to the interpolated map [36], and are given by:

$$\lambda^G(x_i) = \frac{1}{N_B} \sum_{B=1}^{N_B} \lambda_i^B(x_i) \tag{4}$$

where $N_B$ is the total estimated number of blocks and $x_i$ is the location of the $i$th sampled well.

The global kriging weights provide relative rankings of well locations in terms of independent spatial information which gives a logical interpretation that the wells with the lowest global kriging weights are predominantly spatially redundant wells. If the temporary removal of the subsets of these wells does not significantly affect the nature of plume map, these wells were permanently removed. This process is repeated until the removal of a subset of wells change the plume map. Performing this process subsequently creates a situation where the removal of subsets significantly changes the plume. At this point in time, the subset is not removed from the network. In order to fix the limitation for such changes in plume map, a two-sided (upper and lower) confidence interval of 95% is assigned while limiting median plume concentration.

Again, the relative spatial uncertainty for the installation of new monitoring wells in the existing LTM network based on the local kriging variance is given by the global kriging variance ($kv^G$), defined as:

$$kv^{G} = \frac{1}{N_B} \sum_{B=1}^{N_B} kv^{B}(x_B) \tag{5}$$

where $x_B$ denotes the location of the *B*th block and $kv^B$ ($x_B$) is the local kriging variance of the *B*th block [36].

The groundwater LTM network has been spatially optimized for MCB, α-HCH and $SO_4^{2-}$ in 2-D and 2.5-D of the groundwater aquifer. The inherent assumption in 2-D analysis is the existence of all well locations in flat 2-D which is most applicable with a single, fairly uniform and well-connected aquifer. In contrast, the 2.5-D analysis assumes multiple hydrogeological layers in the aquifers without hydraulic interconnection in which LTM is optimized for separate hydrogeological layers. The Quadratic Logistic Regression (QLR) mapping algorithm then uses the data (segregated into subsets representing one chemical of concern for each vertical zone and time slice triplet) from a given subset to map the layer and time frame represented by that given triplet.

### 3.2.2. Temporal Optimization Method

To optimize groundwater monitoring network temporally, Sen's method [38] was applied which is based on iterative thinning. The algorithm consists of (i) estimating a trend using the entire time series; (ii) thinning the time series by a randomly selected fraction of the measurements; and then (iii) re-estimating the trend to determine if the slope estimate is still close to the original slope. The random removal is only retained if the measured slope is within the limits of the confidence interval on the slope using the full dataset [36]. To eliminate the effect of statistical assumptions inherent in standard linear regression methods, trend estimation is carried out using Sen's method [38,39]. Sen's procedure is non-parametric, readily adapted to non-detect measurements and irregular sampling frequencies [40]. Furthermore, this method precludes the calculation of slope for missing points and can predict the median slope even if the number of undetected measurements is less than $(n - 1)/2$.

Sen's estimate (Q) is simply the median value of the resulting list of slopes and is given by:

$$\mathrm{Q} = \frac{x_j - x_i}{j - i} \tag{6}$$

$Q$ is the slope between data points; $x_i$ and $x_j$ are concentrations measured at times $i$ and $j$; Time $j$ is after time $i$ ($j > i$).

$$\begin{aligned} Q &= \text{median slope} = Q_{[(N+1)/2]} \text{ if } N \text{ is odd} \\ &= (Q_{[N/2]} + Q_{[(N+2)/2]})/2 \text{ if N is even} \end{aligned} \tag{7}$$

where $N$ is the number of calculated slopes

A two side ($M_1$, upper and $M_2$, lower) confidence interval for the median slope is estimated using $Z_{statistic}$ and Mann-Kendall statistic ($VAR(S)$). If there is a two-sided confidence interval of 95%, the $Z_{(1-0.05/2)} = Z_{0.975} = 1.96$. Mann-Kendall statistic ($VAR(S)$) [41,42] is given by:

$$VAR(S) = \frac{1}{18}\left[n(n-1)(2n+5) - \sum_{p-1}^{q} t_p(t_p - 1)(2t_p + 5)\right] \tag{8}$$

where $n$ is number of sampling data points, $t_p$ is the number of ties for the $p$th value and $q$ is the number of tied values. Equation (8) may be used for values between 10 and 40. The range of ranks for the specified confidence interval ($C_i$) [38] is given as shown below:

$$C_i = Z_{1-i/2} \times \sqrt{VAR(S)} \tag{9}$$

Taking the value of Equation (9), the ranks of the lower ($M_1$) and upper ($M_2 + 1$) confidence limits can be calculated using the following relation:

$$M_1 = \frac{N - C_i}{2} \text{ And } M_2 = \frac{N + C_i}{2} \tag{10}$$

The values of lower ($M_1$) and upper ($M_2 + 1$) confidence limits were used to define lower and upper boundaries along the median slope. The temporal optimization of the existing groundwater monitoring network was carried out using α-HCH, MCB and $SO_4^{2-}$ concentration data set from 2003–2009 for each contaminant separately and together.

## 3.3. Factors Influencing Monitoring Optimization Results

### 3.3.1. Block Width and Dimension Dependency

In the monitoring network optimization using geostatistical methods, a plume map of well location is created with the help of global kriging weight [8] computed by interpolation method which depends upon width and number of blocks. Hence, in the optimization process, a block width from 1000 m to 1 m was demarcated in order to discover ambiguities in the method.

The 2-D treatment ignores the sampling depth of monitoring well while in the 2.5-D case, the aquifer is assumed to have the number of hydrogeological layers and thus the sampling depth of the monitoring wells needs to be taken into account. For systematic study of the influence of dimension and block width on monitoring network optimization methods, the optimization process has considered block widths from 1000 m to 1 m in both 2 and 2.5 dimensions.

### 3.3.2. Contaminants Association

Based on PCA of hydro-geochemical data, contaminant concentration of three representative variables, α-HCH, MCB and $SO_4^{2-}$ are used individually to monitor network optimization. The likely influence of multiple variables on optimization process is examined using data sets in groups as well as all three variables together. In this case, co-kriging was used to compute co-kriging numerical weights from contaminant concentrations for each monitoring well which in turn were used to create a plume map of the monitoring network which is used for geostatistical optimization of monitoring network as explained in Section 3.2.1.

## 3.4. Groundwater Flow Direction and Aquifer Homogeneity

In addition to spatio-temporal optimization of ground water monitoring network, direction of ground-water flow and homogeneity of the aquifer was determined. Groundwater flow direction is known to play a prominent role in monitoring network optimization. The directional dependency was studied using directional variograms. Estimation of experimental variogram was carried out using Golden Surfer software [43].

Since variograms based on contaminant concentration data set were anisotropic in nature, valid variogram model incorporating directional dependence were constructed using standard models to model the data set. The best fitting model, considering range, sill, nugget effect and shape of the model, was selected. A variogram was built from a concentration data set of α-HCH, MCB and $SO_4^{2-}$ for the years 2003–2009, according to seasons (summer: May–October; winter: November–April) and to hydrologic seasons (March–May: High groundwater level, September–November: Low groundwater level). The range and sill were estimated for various directional variogram models.

A homogeneity index (RV index) which numerically estimates homogeneity of the aquifer was defined based on range, sill and variance of the variogram [44] based on the variogram model as given by:

$$\text{Homogeneity index} = \frac{\text{range}}{\frac{sill + variance}{2}} \tag{11}$$

## 3.5. Integrating Approaches and Improving Groundwater Monitoring Strategy

New sets of methods were analyzed with new approaches based on integrated statistical and geostatistical methods, integrated in the light of various optimization objectives. However, basic framework and components of groundwater monitoring strategies were duly taken into consideration while subjecting these methods to integration. The new integrated methods, although based on different approaches, could be used as an impeccable tool for groundwater monitoring network optimization and feedback for updating groundwater monitoring strategies. While integrating these methods, the advantages and disadvantages of the methods being used have also been analysed.

# 4. RESULTS

## 4.1. Representative Variable in the Research Area

For the purpose of dimension reduction and analysis of major components loading the system, PCA was applied to a matrix of hydro-geochemical data. The first four principal components in this dataset with eigenvalues greater than one explain 66% of the cumulative variance. The significance of the parameters in four components was further depicted by the component matrix extracted with PCA and varimax with the Kaiser Normalization rotation method. At 75% significance level, $SO_4^{2-}$, Fe, pH and Eh play a major role in the system whereas α-HCH and MCB are minor components. In the first component, $SO_4^{2-}$ and Fe have positive loading. In the second component, Eh has positive loading. However, pH is has negative loading because the acidic groundwater environment in the study area arises from a high $SO_4^{2-}$ concentration. $SO_4^{2-}$ and Fe have positive loading in the first component whereby Eh has positive loading in the second component. The negative loading of pH could be linked to the acidic groundwater environment in the study area from high $SO_4^{2-}$-concentration.

## 4.2. Optimization of LTM Network

### 4.2.1. Spatial Optimization of LTM Network

Both types of aquifers were considered separately during the spatial optimization of LTM network in accordance with the methods described in Section 3.2. The optimized monitoring network suggests 292 wells out of 462 wells in the Quaternary aquifer and 256 wells out of 357 wells in the Tertiary aquifer for monitoring needs at the suggested temporal interval.

The spatial uncertainties analysis, resulting from the global kriging variance (using Equation (5)) made recommendations for 22 and 41 new monitoring wells, respectively, in the Tertiary and Quaternary aquifers. The spatial distribution of redundant, essential, and proposed new wells in the Quaternary and Tertiary aquifer is depicted in Figure 4. Similar kriging weights in the original and optimized LTM network datasets points out that reduction in observation points does not compromise the quality or resolution of the collected samples as long as the distribution is logically and conveniently designed.

### 4.2.2. Temporal Optimization of LTM Network

Monitoring network was temporally optimized through statistical approach using Sen's method, with 95% confidence intervals around the slope estimates, as described in Section 3.2.2. The outputs from temporal optimization are the sampling intervals at 3 quartiles (lower, median and upper) for individual monitoring well and each monitoring parameter (Table 2). The optimization results reveal that the optimization differs remarkably while considering pair and multiple contaminants.

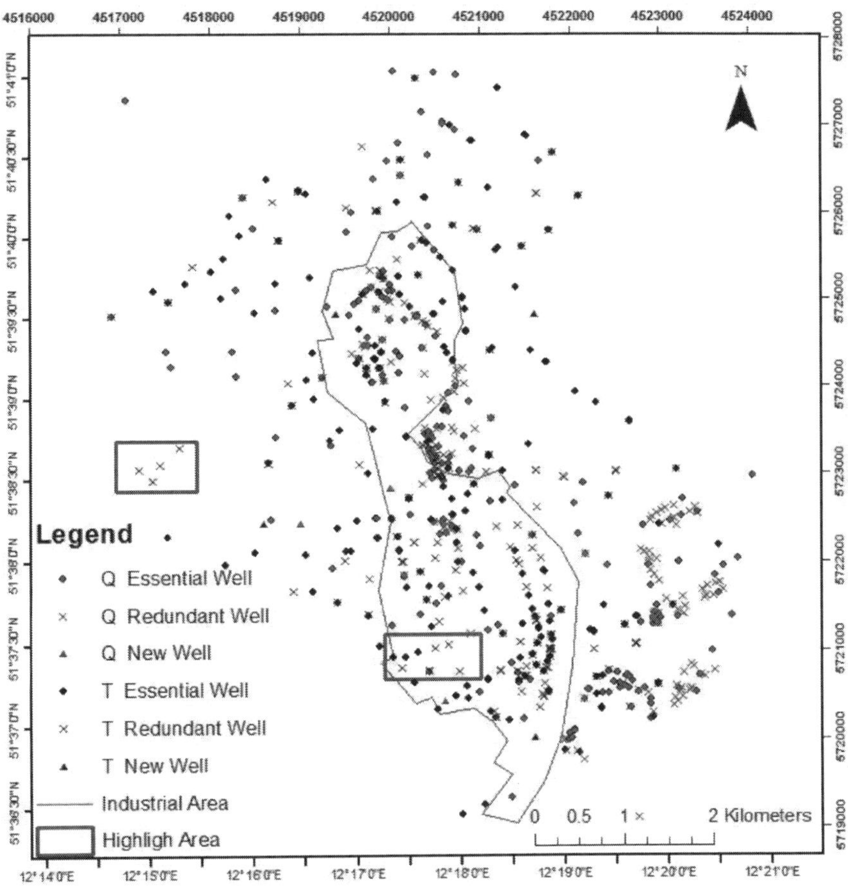


**Figure 4.** Optimized LTM network map showing location of essential, redundant, and proposed new wells in the monitoring network.

In order to clearly visualize and precisely describe the temporal optimization results of three contaminants in combination, the recommended median quartile sampling frequency for the monitoring wells was divided into five classes: three months, six months, one year, two years and three years. Recommendations for specific number of monitoring wells with respect to each of the categorized temporal sampling interval are specified in Table 3.

**Table 2.** Temporal optimization of monitoring network in Quaternary and Tertiary aquifer for monochlorobenzene (MCB), α-HCH and $SO_4^{2-}$.

| V. Zone | COC | Present Sampling Interval (Days) | | | Recommended Sampling Interval (Days) | | |
|---|---|---|---|---|---|---|---|
| | | Lower Quartile | Median Quartile | Upper Quartile | Lower Quartile | Median Quartile | Upper Quartile |
| Q | MCB | 92 | 210 | 254 | 205 | 328 | 503 |
| Q | α-HCH | 138 | 181 | 224 | 298 | 327 | 418 |
| Q | $SO_4^{2-}$ | 179 | 217 | 268 | 325 | 506 | 704 |
| T | MCB | 143 | 221 | 280 | 282 | 443 | 810 |
| T | α-HCH | 224 | 224 | 224 | 342 | 398 | 553 |
| T | $SO_4^{2-}$ | 188 | 217 | 278 | 435 | 589 | 836 |
| Q | all | 138 | 210 | 254 | 298 | 328 | 503 |
| T | all | 188 | 221 | 278 | 342 | 443 | 810 |
| Both | all | 161 | 217 | 261 | 311 | 420 | 628 |

Note: Q: Quaternary aquifer; T: Tertiary aquifer; COC: Chemical of Concern.

**Table 3.** Number of monitoring wells for each sampling interval.

| Sampling Interval | 3 Months | 6 Months | 1 Year | 2 Years | 3 Years | Total |
|---|---|---|---|---|---|---|
| Quaternary | 34 | 86 | 173 | 76 | 93 | 462 |
| Tertiary | 16 | 69 | 114 | 84 | 74 | 357 |
| Total No. of wells | 50 | 155 | 287 | 160 | 167 | 819 |

Figure 5 shows the distribution of monitoring wells and their recommended sampling intervals. The highest number of sampling wells is recommended at the yearly sampling interval.

In the study area, the overall temporally optimized sampling interval was recommended as 238 days, 317 days and 401 days in terms of lower, median and upper quartiles respectively.

## 4.3. Factor Influencing Monitoring Optimization Results

### 4.3.1. Dimension and Block Width Dependency

With an objective of incorporating dependency of dimension and block width on optimization process, the optimization was carried out for different interpolation block widths (from 1000 m to 1 m) for both aquifers in both 2 and 2.5 D dimensions as described in Section 3.3.1.

***Optimization in 2-D Aquifer***

Considering the aquifer as a 2-D plain, groundwater monitoring network was spatially optimized for MCB, α-HCH and $SO_4{}^2$. The result unveiled a significant change in number of suggested/essential wells as well as existing redundant wells with changing block width for interpolation (from 1000 m to 1 m). The increase in the relative spatial uncertainties based on local kriging variance with decreasing block width eventually recommends installation of new monitoring wells.

The overall analysis of optimization in 2-D plain unveils that the relative spatial uncertainty increases with decreasing block width from 200 m to 1 m in both aquifers (Figure 6 and Figure 7), thus suggesting the need of installation of 63 and 36 new wells, respectively, in Quaternary and Tertiary aquifers at 1 m block width.

***Optimization in a 2.5-D Aquifer***

Spatial optimization for MCB, α-HCH and $SO_4{}^{2-}$ in groundwater LTM network has been conducted in combination in 2.5-D aquifer. The LTM network is optimized separately for each hydro-stratigraphic layer in the aquifer using data from that layer only.

In the case when all three contaminants are considered together, the change in block width for interpolation (1000 m to 1 m) does not significantly change the number of suggested wells (Figure 8 and Figure 9). The increment of spatial relative uncertainty significantly with decreasing block width from 20 m to 1 m suggests the installation of 58 and 38 new monitoring wells in Quaternary and Tertiary aquifers, respectively.

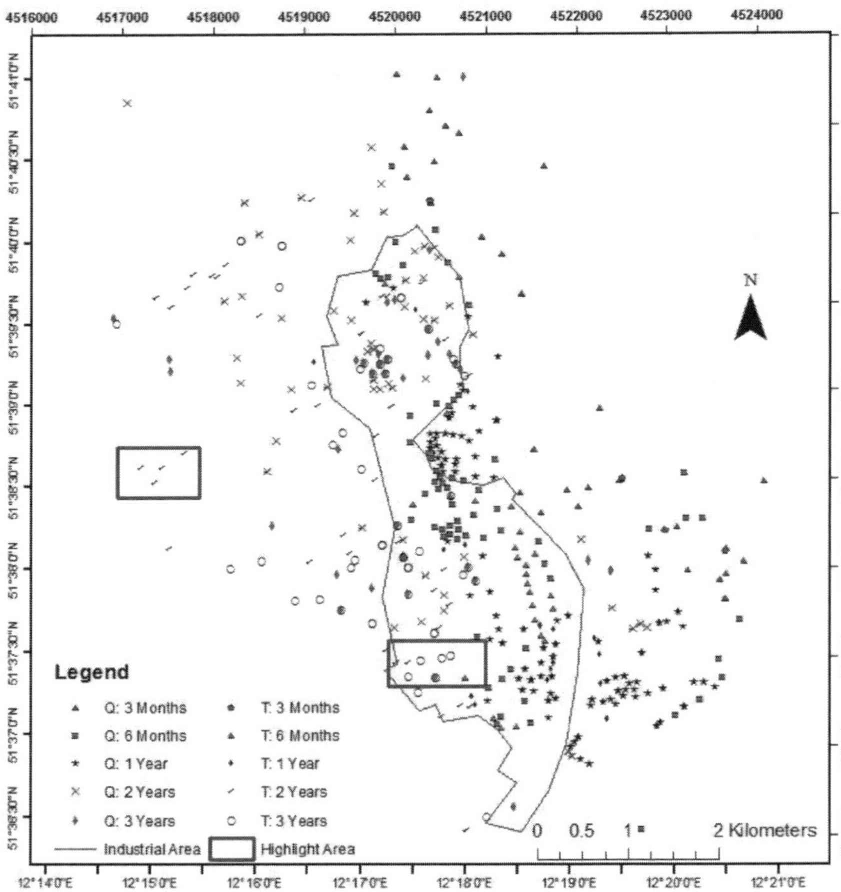


**Figure 5.** Statistically temporally optimized LTM network map showing recommended temporal frequency of the monitoring wells in the monitoring network.

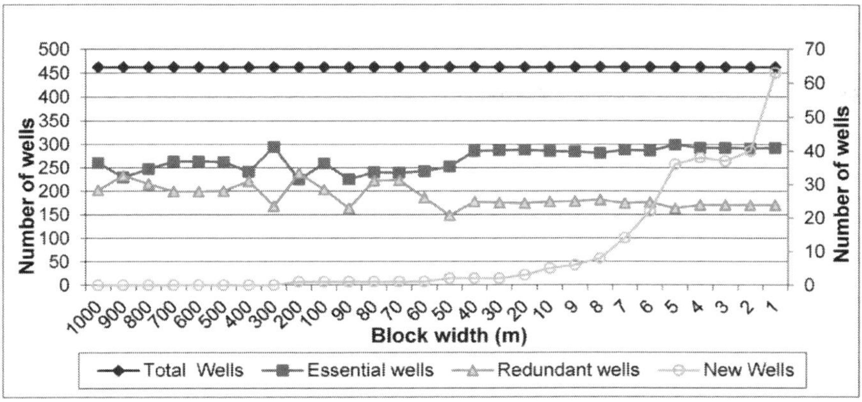


**Figure 6.** Block width dependency in the LTM network optimization in the Quaternary aquifer for MCB, α-HCH and $SO_4^{2-}$ in a 2-D plain. Essential, redundant and total number of wells is on left Y-axis and number of new wells is on right Y-axis.

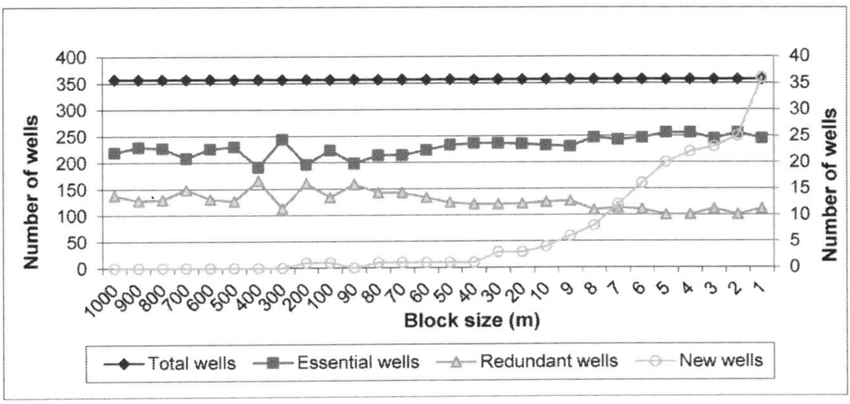


**Figure 7.** Block width dependency in the LTM network optimization in the Tertiary aquifer for MCB, α-HCH and $SO_4^{2-}$ in a 2-D plain. Essential, redundant and total number of wells are on left Y-axis and the number of new wells is on right Y-axis.

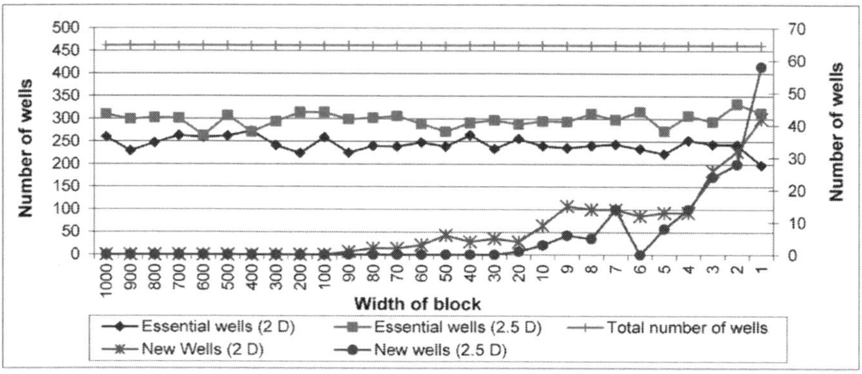


**Figure 8.** Block width dependency in the LTM network optimization in the Quaternary aquifer for MCB, α-HCH and $SO_4^{2-}$ considering 2-D and 2.5-D aquifers. Essential, redundant and total number of wells are on left Y-axis and number of new well is on right Y-axis.

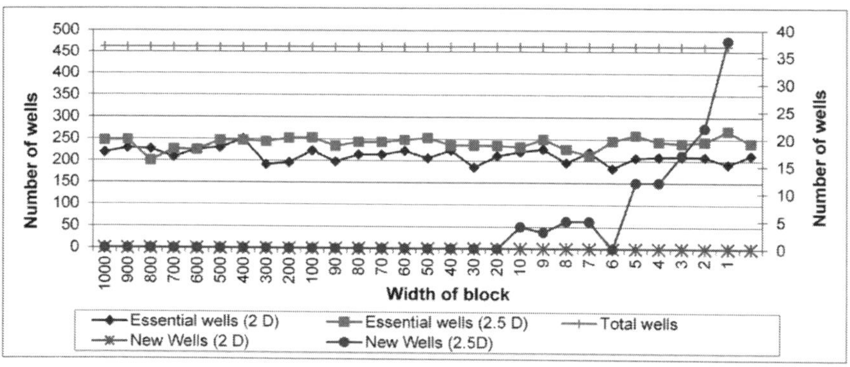


**Figure 9.** Block width dependency in the LTM network optimization in the Tertiary aquifer for MCB, α-HCH and $SO_4^{2-}$ in 2-D and 2.5-D aquifers. Essential, redundant and total number of wells are on the left Y-axis and the number of new wells is on the right Y-axis.

### 4.3.2. Contaminants Association

To account for the association between the contaminants, the LTM network was spatially optimized taking 7 m interpolation block width for the contaminants MCB, α-HCH and $SO_4^{2-}$ in pairs as well as all grouped together, considering the groundwater aquifer separately as 2-D plain and as 2.5-D aquifer. Dependency of contaminants association on network optimization is supported by result that the spatial optimizations of the LTM network for MCB and α-HCH individually show less redundancy and low recommendations for new monitoring wells while that only for $SO_4^{2-}$ recommends higher number of redundant wells (70%). Such effect of incorporating greater number of contaminants is presented in Table 4 andTable 5. For example, in the Quaternary aquifer considered as a 2-D plain, for α-HCH and $SO4^{2-}$, 73 wells are recommended, for MCB and $SO_4^2$, 293 wells are recommended and for MCB and α-HCH, 335 wells are recommended. However, 244 wells are recommended when considering all three contaminants MCB, $SO_4^{2-}$ and α-HCH together (Table 4). Similarly, in the tertiary aquifer considered as a 2.5-D plain, 72, 267, 247 and 213 wells are recommended, respectively, for combination of α-HCH and $SO_4^{2-}$; MCB and $SO_4^{2-}$; MCB and α-HCH; and MCB, $SO_4^{2-}$ and α-HCH together (Table 5).

In the optimization of LTM network for combination of three contaminants, the local kriging weight of each contaminant is averaged, which clearly depicts the dependence of relative spatial uncertainty with the spatial distribution of each contaminants. It has been observed that optimization for each type of aquifer considering individual contaminants and in groups gives recommendations for different numbers and locations of new wells (Table 4 and Table 5).

## 4.4. Groundwater Contaminants Flow Direction and Aquifer Homogeneity

Geostatistical and flow direction modeling methods have been used to predict the groundwater flow direction and integrate its effect into LTM network optimization. The spatial characterization of the groundwater contamination scenario (degree of spatial co-relation between contaminant concentration values) was observed using an experimental variogram exhibiting the contaminant concentration data of MCB, α-HCH and $SO_4^{2-}$.

**Table 4.** Number of essential, redundant and new well location recommendation from 2-D groundwater monitoring network optimization for α-HCH and $SO_4^2$; MCB and $SO_4^{2-}$; MCB and α-HCH; and MCB, $SO_4^{2-}$ and α-HCH in Quaternary (462 wells) and Tertiary (357 wells) aquifers.

| **Quaternary Aquifer/Contaminants** | **α-HCH and $SO_4^{2-}$** | **MCB and $SO_4^{2-}$** | **MCB and α-HCH** | **MCB, $SO_4^{2-}$ and α-HCH** |
|---|---|---|---|---|
| Essential well | 73 | 293 | 335 | 244 |
| Redundant well | 351 | 169 | 127 | 218 |
| New well location | 14 | 7 | 14 | 14 |
| **Tertiary Aquifer** | | | | |
| Essential wells | 52 | 233 | 247 | 220 |
| Redundant well | 302 | 124 | 110 | 137 |
| New well location | 0 | 0 | 0 | 0 |

**Table 5.** Number of essential, redundant and new well location recommendation from 2.5-D groundwater monitoring network optimization for α-HCH and $SO_4^2$; MCB and $SO_4^{2-}$ MCB and α-HCH; and MCB, $SO_4^{2-}$ and α-HCH in Quaternary (462 wells) and Tertiary (357 wells) aquifers.

| **Quaternary Aquifer/Contaminants** | **α-HCH and $SO_4^{2-}$** | **MCB and $SO_4^{2-}$** | **MCB and α-HCH** | **MCB, $SO_4^{2-}$ and α-HCH** |
|---|---|---|---|---|
| Essential well | 107 | 334 | 341 | 298 |
| Redundant well | 317 | 127 | 140 | 164 |
| New well location | 0 | 2 | 1 | 14 |
| **Tertiary Aquifer** | | | | |
| Essential well | 72 | 267 | 247 | 213 |
| Redundant well | 282 | 90 | 148 | 144 |
| New well location | 0 | 3 | 8 | 5 |

The geometric anisotrophy as depicted by the experimental variogram was observed for each 30° of lag direction with reference to North direction. Again, the heterogeneity in the nature of geological formation was quantified by using a homogeneity index called RV Index (Equation (11)), which takes its base from spatial variability of contaminant concentration distribution.

### 4.4.1. Contaminant Wise Variogram Modelling

Based on the results from experimental variogram, geometrical anisotropy was observed in the data set of α-HCH, as range and sill differed in different directions.

The experimental variogram (incorporating each 30° of lag direction) revealed the highest range in Northern direction (0°) which is also coherent with the results from RV index that was highest in the Northern direction for Quaternary aquifer (Table 6). Meanwhile, the variogram showed highest range and RV index at 30° from the Northern direction for the Tertiary aquifer (Table 6). These results specify that the overall conspicuous concentration of α-HCH flowed towards the North in both aquifers. In case of Tertiary aquifer, there was slight deviation in the flow towards east direction.

The relatively lower RV index in the Quaternary aquifer than in the Tertiary aquifer shows high heterogeneity in the Quaternary aquifer.

### 4.4.2. Season Wise Variogram Modelling for α-HCH Data

Seasonal variability was embodied into optimization by using the data set of α-HCH for hydrological summer and winter seasons to model the experimental variogram. Periodic data from May to October (2003–2009) and November to April (2003–2009) were categorized as hydrological summer and winter seasons, respectively. Analysis of the experimental variogram, using α-HCH data for the hydrological summer season, again showed the highest range in the North direction as well as highest RV index in Northern direction for both Quaternary and Tertiary aquifers (Table 7 and Table 8). These results show that preferential groundwater flow and α-HCH concentration movement occurred in the Northern direction.

**Table 6.** Directional variogram modelling α-HCH concentration in the long term monitoring network (LTM) network in Quaternary and Tertiary aquifers (January 2003 to February 2009).

| α-HCH for Quaternary Aquifer (2003–2009) | | | | α-HCH for Tertiary Aquifer (2003–2009) | | | |
|---|---|---|---|---|---|---|---|
| Model: Gaussian, Variance: 0.62 | | | | Model: Gaussian, Variance: 0.57 | | | |
| Direction | Range | Sill | RV Index | Direction | Range | Sill | RV Index |
| Omni | 404 | 0.64 | 641.27 | Omni | 1182 | 0.70 | 1862.88 |
| 0 | 666 | 0.62 | 1074.19 | 0 | 960 | 0.79 | 1416.97 |
| 30 | 464 | 0.64 | 736.51 | 30 | 1340 | 0.67 | 2161.29 |
| 60 | 385 | 0.66 | 601.56 | 60 | 999 | 0.69 | 1591.91 |
| 90 | 321 | 0.66 | 501.56 | 90 | 710 | 0.70 | 1118.11 |
| 120 | 360 | 0.65 | 566.93 | 120 | 722 | 0.70 | 1137.01 |
| 150 | 433 | 0.63 | 692.80 | 150 | 750 | 0.68 | 1200.00 |
| 180 | 590 | 0.62 | 951.61 | 180 | 960 | 0.69 | 1529.76 |

In the next round, the experimental variogram was modelled using α-HCH data for the hydrological winter season (November to April, 2003–2009). The experimental variogram showed the highest range in the direction of 60° from North in the Quaternary aquifer (Table 8) whereas the range was highest in the direction of 120° from North in Tertiary aquifer. Further, the RV index is lower in the Quaternary aquifer than the Tertiary aquifer, thus indicating high heterogeneity in Quaternary aquifer (Table 8).

**Table 7.** Directional variogram modelling for α-HCH concentration in the LTM network in Quaternary aquifers for summer and winter seasons (May to October in 2003–2009).

| α-HCH for Quaternary Aquifer (May to October in 2003–2009) | | | | α-HCH for Quaternary Aquifer in Winter Seasons (Nov. to April from 2003–2009) | | | |
|---|---|---|---|---|---|---|---|
| Model: Gaussian, Variance: 0.70 | | | | Model: Gaussian, Variance: 0.68 | | | |
| Direction | Range | Sill | RV Index | Direction | Range | Sill | RV Index |
| Omni | 491 | 0.78 | 661.99 | Omni | 669 | 0.73 | 954.01 |
| 0 | 491 | 0.70 | 701.43 | 0 | 398 | 0.74 | 562.54 |
| 30 | 387 | 0.75 | 532.54 | 30 | 534 | 0.71 | 771.12 |
| 60 | 340 | 0.77 | 461.52 | 60 | 543 | 0.68 | 801.48 |
| 90 | 354 | 0.78 | 477.28 | 90 | 397 | 0.72 | 570.20 |
| 120 | 364 | 0.79 | 488.59 | 120 | 411 | 0.73 | 585.47 |
| 150 | 333 | 0.76 | 456.16 | 150 | 422 | 0.74 | 596.89 |
| 180 | 390 | 0.65 | 577.78 | 180 | 513 | 0.73 | 729.21 |

**Table 8.** Directional variogram modelling of α-HCH concentration in the LTM network in Quaternary and Tertiary aquifers for winter seasons (November to April from 2003–2009).

| α-HCH for Tertiary Aquifer (May to October in 2003–2009) | | | | α-HCH for Tertiary Aquifer in Winter Seasons (Nov. to April from 2003–2009) | | | |
|---|---|---|---|---|---|---|---|
| Model: Gaussian, Variance: 0.70 | | | | Model: Gaussian, Variance: 1.25 | | | |
| Direction | Range | Sill | RV Index | Direction | Range | Sill | RV Index |
| Omni | 1166 | 0.41 | 2854.69 | Omni | 585 | 1.41 | 439.85 |
| 0 | 1258 | 0.47 | 2875.43 | 0 | 579 | 1.31 | 452.34 |
| 30 | 1062 | 0.46 | 2455.49 | 30 | 550 | 1.25 | 440.00 |
| 60 | 939 | 0.38 | 2392.36 | 60 | 600 | 1.33 | 465.12 |
| 90 | 888 | 0.43 | 2122.12 | 90 | 850 | 1.26 | 677.29 |
| 120 | 737 | 0.43 | 1761.26 | 120 | 1050 | 1.22 | 850.20 |
| 150 | 793 | 0.42 | 1934.15 | 150 | 823 | 1.25 | 658.40 |
| 180 | 882 | 0.44 | 2082.89 | 180 | 526 | 1.32 | 409.34 |

### 4.4.3. Year Wise Directional Variogram Modelling for α-HCH Data

To acquaint with the geometrical anisotropy of the data set, variogram modelling was carried out for the monitoring data sets of the years 2005 and 2006. To understand the year wise alterations in the homogeneity of aquifer and flow direction, variogram modelling was carried out for the monitoring data sets of the years 2005 and 2006. The experimental variogram showed that the range was highest in the Northern direction, as shown in Table 9 and Table 10.

Again, the range was highest at 30° from North in the Tertiary aquifer. The RV index varied with different lag directions, showing no distinct pattern or peak.

**Table 9.** Directional variogram modelling α-HCH concentration in the LTM Network in Quaternary and Tertiary aquifers for 2005.

| α-HCH for Quaternary Aquifer in 2005 | | | | α-HCH for Tertiary Aquifer in 2005 | | | |
|---|---|---|---|---|---|---|---|
| Model: Gaussian, Variance: 0.89 | | | | Model: Gaussian, Variance: 1.05 | | | |
| **Direction** | **Range** | **Sill** | **RV Index** | **Direction** | **Range** | **Sill** | **Index** |
| Omni | 850 | 0.89 | 955.06 | Omni | 750 | 1.33 | 631.31 |
| 0 | 500 | 0.93 | 550.96 | 0 | 937 | 1.30 | 797.45 |
| 30 | 533 | 0.92 | 588.30 | 30 | 1125 | 1.35 | 937.50 |
| 60 | 758 | 0.95 | 826.16 | 60 | 933 | 1.30 | 794.04 |
| 90 | 856 | 0.93 | 939.63 | 90 | 750 | 1.30 | 638.30 |
| 120 | 800 | 0.94 | 874.32 | 120 | 750 | 1.30 | 638.30 |
| 150 | 533 | 0.95 | 580.93 | 150 | 562 | 1.32 | 474.26 |
| 180 | 500 | 0.91 | 555.56 | 180 | 687 | 1.33 | 577.31 |

## 4.5. Integrating Approaches for Improving Groundwater Monitoring

As an important objective of this research, statistical and geostatistical methods were taken as a basis for integrating new approaches to illustrate several sets of methods for different optimization objectives. Figure 10 depicts integration of statistical and geostatistical methods. Such integration was executed to fully understand, evaluate and optimize groundwater monitoring with reference to groundwater monitoring of a contaminated site.

**Table 10.** Directional variogram modelling α-HCH concentration in the LTM Network in Quaternary and Tertiary aquifers for 2006.

| α-HCH for Quaternary Aquifer in 2006 | | | | α-HCH for Tertiary Aquifer in 2006 | | | |
|---|---|---|---|---|---|---|---|
| Model: Gaussian, Variance: 0.59 | | | | Model: Gaussian, Variance: 1.05 | | | |
| **Direction** | **Range** | **Sill** | **RV Index** | **Direction** | **Range** | **Sill** | **RV Index** |
| Omni | 375 | 0.65 | 605.23 | Omni | 1333 | 0.81 | 1904.29 |
| 0 | 468 | 0.61 | 793.22 | 0 | 1684 | 0.72 | 2405.71 |
| 30 | 437 | 0.68 | 740.68 | 30 | 1727 | 0.84 | 2467.14 |
| 60 | 375 | 0.67 | 635.59 | 60 | 1333 | 0.86 | 1904.29 |
| 90 | 375 | 0.67 | 635.59 | 90 | 1058 | 0.84 | 1511.43 |
| 120 | 375 | 0.66 | 635.59 | 120 | 1058 | 0.84 | 1511.43 |
| 150 | 375 | 0.66 | 635.59 | 150 | 1055 | 0.68 | 1507.14 |
| 180 | 375 | 0.53 | 635.59 | 180 | 1277 | 0.69 | 1824.29 |

Multivariate and descriptive statistics were the primary element of groundwater monitoring framework used for analysis, classification, modelling and interpretation of large volume of available data set. The second element consisted of optimization of existing LTM networks based on target variables (MCB, α-HCH and $SO_4^{2-}$) using a geostatistical spatial optimization algorithm (GTA). GTS itself accounted for dimension dependency, influence of block width for interpolation and the influence of multiple contaminants on the LTM network optimization, whereas variogram modelling estimated the groundwater flow direction and heterogeneity of aquifers. As third and fourth elements of the groundwater monitoring framework, the obtained results were analysed and interpreted incorporating the prominent effects of other influencing factors such as legal requirements and land use changes so that essential, redundant and new monitoring wells in the existing LTM network could be recommended.

Strategic planning of monitoring network optimization is crucial to groundwater monitoring since it is an important component of monitoring strategies [45,46]. With an overall objective to improve the monitoring strategies, the developed and existing methods were applied to the observed and model-based data sets for the mega-contaminated site, Bitterfeld/Wolfen.

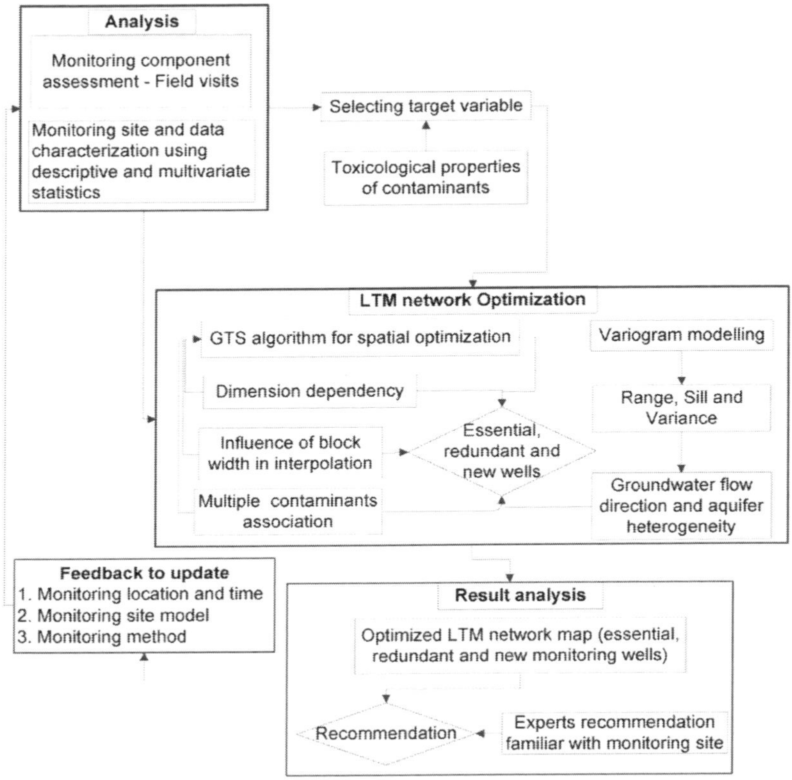


**Figure 10.** Research steps showing the integration of statistical and geostatistical methods for groundwater monitoring.

The existing monitoring strategies can be improved by incorporating the optimization outcomes from integrated methods along with functionality consideration of expert knowledge pertaining to the study area.

# 5. DISCUSSION

PCA technique is an effective technique for pattern recognition which attempts to explain the variance of a data set of intercorrelated variables with small set of independent variables [47]. Representative variables for groups of groundwater contaminants ($SO_4^{2-}$, α-HCH and MCB) were selected for monitoring network optimization in the study area based on these PCA results.

The optimization results from geostatistical methods suggests the need for monitoring for only 292 out of 462 wells in the Quaternary aquifer and 256 out of 357 wells in the Tertiary aquifer (Figure 4). Furthermore, the spatial uncertainty analysis in terms of kriging variance recommends 22 and 41 new monitoring wells to be added to the existing network in the Tertiary and the Quaternary aquifers, respectively, when three representative contaminants (α-HCH, MCB and $SO_4^{2-}$) are considered together. Given the presence of variety of highly mobile and persistent contaminants in the research area [48,49], monitoring network optimization with more number of available contaminants would result in better optimization (as presented inSection 3.3.2. and Section 4.3.2).

Temporal optimization was performed with due consideration of pairs of contaminants as well as three contaminants together; recording the intervals in terms of lower quartile, median quartile and upper quartile for each monitoring well for each parameter. The result shows that the optimization differs markedly for single and multiple contaminants (Table 2).

Optimization in the 2-D plain considering three contaminants together shows the increase in relative spatial uncertainty with decreasing block width form 200 m to 1 m in both aquifers (Figure 6 and Figure 7) suggesting installation of 63 and 36 new wells, respectively, in Quaternary and Tertiary aquifers at 1 m block width. Again, with the same consideration of contaminants in 2.5-D analysis, the change in block width for interpolation (1000 m to 1 m) has an insignificant effect on the number of suggested wells (Figure 8 and Figure 9). In this case, the increment of spatial relative uncertainty significantly with decreasing block width from 20 m to 1 m suggests the installation of 58 and 38 new monitoring wells in Quaternary and Tertiary aquifers, respectively. A 2-D analysis treats all wells in a manner that they exist in a flat 2-D plane which should be more applicable in case of single, fairly uniform and well-connected aquifer [50]. On the other hand, a 2.5-D analysis treats the aquifer as comprising multiple Hydrogeological layers, each of which is considered separately during optimization within the GTS [36]. Again, the analysis reveals notable dependence of optimization of existing monitoring network on block width. With the gradual decrease in the block width, the recommendation of the number of new monitoring wells increase. The incorporation of the relatively influential role of block width for interpolation has not been documented in previous studies based on geostatistical methods [13,16]. Hence, this study has pioneered the incorporation of this factor in the monitoring network optimization process. Previous studies have, however, addressed problems related to scale in hydrology such as the works of Bergström and Graham [51], Dooge [52], Gupta *et al.* [53], Sivapalan and Kalma [54] and Sivapalan *et al.* [55]. The results based on the present study gives recommendations of scale size of 100 m to 7 m for monitoring network optimization at local scale.

The role of contaminants association was pronounced when three representative contaminants were considered in different pairs or all together (Table 4 and Table 5). The optimization result in terms of the number of essential, redundant and new wells was found to be different. Spatial variograms for each contaminant was first estimated for the spatial covariance model, which was determined through a non-linear modelling program. This modelling utilized the Levenberg-Marquardt algorithm [56,57]. The fitting algorithm was set up to fit either a combination of up-to-three spherical, exponential and/or Gaussian components or a combination of up-to-three power model structures [36]. As the spatial variogram for each contaminant varies, the combination of two or more contaminants determines the covariance of the contaminants for the kriging interpolation.

The analysis of flow direction (from January 2003 to February 2009) shows that α-HCH flows northwards in the Quaternary aquifer, whereas in the direction 30° from north in the Tertiary aquifer. This distinction in the flow direction with respect to aquifer type is the result of historic shift in flow direction from northwards to eastwards and then back to northwards. This represents source specific spreading direction of the α-HCH contaminant. The seasonal alterations in the groundwater flow system has also been documented in previous studies [58,59]. Seasonal influence was observed in both aquifers since the experimental variogram modelling using the α-HCH data shows north as the preferential flow direction in hydrological summer season (May to October in 2003–2009) while during hydrological winter season, the flow direction was 60° from north in the Quaternary aquifer and 120° from north in the Tertiary aquifer (Table 8). Seasonal fluctuation in the water level in Mulde river and surrounding water bodies [60] and the location of contaminant source strongly control such seasonal variation in the flow direction.

The geometric anisotropy in contaminant concentration data set using experimental variogram was the means of determining groundwater and contaminants flow direction (described in Section 3.4). Table 6, Table 7, Table 8 and Table 9show the prominent groundwater and contaminant flow direction that are revealed from the data sets of α-HCH, MCB and $SO_4^{2-}$ concentration. It has been shown that although the prominent groundwater and contaminant flow direction is northwards, various contaminants spread non-uniformly in varied time scale. The RV index points out the hydrogeological homogeneity in the aquifer in the direction where there is the highest range. The likely flow direction and homogeneity analysis in a year wise basis displayed the complete change in flow direction in 2006 which was towards east in Quaternary aquifer, but in the direction of 30° from north in the Tertiary aquifer during 2005. The preferential flow direction was northwards in the Quaternary aquifer and 30° from north in the Tertiary aquifer.

Groundwater monitoring network optimization methods are greatly challenged by other factors besides the problems associated with nature and availability of data. Given the existence of a wide range of methods for LTM network optimization, most of them fail to implicate the factors that influence the optimization methods [7]. New and improved approaches based on integrated statistical and geostatistical methods for optimization of monitoring network along with their possible implications are dealt with in this study. In addition, a range of deterministic factors and their influence on application of these methods in the research area are analyzed. Existing monitoring strategies can be re-shaped by removing the existing redundancy as well as adding requisite newer monitoring wells. A new approach has been devised in this study by integrating statistical and geostatistical methods for spatio-temporal optimization of the LTM network which has been tested in the study area. The outcomes of this research can be taken as a basis for improving the strategies to make the long term monitoring program more efficient and equally effective. However, expert knowledge and site-specific peculiarities are important to take into consideration in order to implement this approach.

# 6. CONCLUSIONS AND RECOMMENDATION

The results have verified that the geostatistical method, based on the kriging interpolation weight, shows realistic optimization results in terms of recommended essential, redundant and new monitoring wells. For the time being, several factors influencing optimization results such as block width, multiple contaminants association, and contaminant spreading direction were analyzed. The results obtained from such analysis revealed their significant role in the monitoring optimization process. Meanwhile, the network was temporally optimized using a statistical approach with the help of Sen's method. Spatio-temporal optimization reduces the number of samples and optimizes the monitoring frequency, which could make the monitoring program more effective in terms of resource requirement. The flexibility inherent in both statistical and geostatistical approaches allow them to be operated at various confidence level limit set by user. Furthermore, these methods could be applied to field based real groundwater quality data set.

Since various methods have strengths and weaknesses, the strength of these methods can be integrated on the basis of optimization objectives of groundwater monitoring strategies. For example, in a monitoring network where there is a low density of monitoring wells, statistical and geostatistical methods can be integrated. An understanding of the distribution of contaminants, component analysis can be gained, as well as temporal optimization can be done with the help of statistical methods, whereas spatial optimization of the monitoring network can be done by using geostatistical methods. Temporal optimization of the monitoring network can be carried out using statistical methods. In this manner, these methods can be systematically integrated in monitoring network optimization according to the prime objective of the groundwater monitoring strategies.

Thus, the study has demonstrated that the existing monitoring network could be agreeably optimized using the presented statistical and geostatistical methods without the obvious fear of losing essential information from the current monitoring network. In order to accomplish sound groundwater resource management practices, necessary improvements in groundwater monitoring strategies is the basic key. As such, the efforts to optimize and evaluate the monitoring network will enhance the performance of the water management system. The methods presented in this research are convenient both in the case of inadequate networks, which need new monitoring wells as well as for dense monitoring networks, with redundant wells. The method can be conveniently used to find redundancy in the existing monitoring network as well as to locate necessary new monitoring wells in developing countries, whereas it is convenient to thin out the dense monitoring network (without losing valuable information) in developed countries. The integrated statistical and geostatistical method can be integrated with expert knowledge and field attributes to improve existing monitoring strategies by adding or removing existing monitoring wells in the monitoring network.

## ACKNOWLEDGMENT

The presented work is funded by a Helmholtz PhD scholarship of the Helmholtz Centre for Environmental Research—UFZ in collaboration with the Martin Luther University Halle-Wittenberg, Germany. Groundwater monitoring data of Bitterfeld/Wolfen was provided by the LAF Sachsen-Anhalt. The data exchange was embedded in the SAFIRA I and II projects (funding number 02WT0023 & 02WT00412). This work has been possible because of kind co-operation and support from the Department of Hydrogeology and Environmental Geology, Institute of Geosciences and Geography, Martin Luther University, Halle (Saale), Germany.

## CONFLICT OF INTEREST

The authors declare no conflicts of interest.

## REFERENCES

1. Wada, Y. Water Stress over the Year: Quantitative Analysis of Seasonality and Severity on a Global Scale. Master's Thesis, Utrecht University, Utrecht, The Netherlands, 2008.
2. Johnson, N.; Revenga, C.; Echeverria, J. Managing water for people and nature. *Science* **2001**, *292*, 1071–1072.
3. Thakur, J.K.; Thakur, R.K.; Ramanathan, A.; Kumar, M.; Singh, S.K. Arsenic contamination of groundwater in nepal—An overview. *Water* **2011**, *3*, 1–20.
4. Van Geer, F.C.; Bierkens, M.F.P.; Broers, H.P. Groundwater monitoring strategies. In *Encyclopedia of Hydrological Sciences*; John Wiley & Sons, Ltd.: Hoboken, NJ, USA, 2006. [
5. Erechtchoukova, M.G.; Chen, S.Y.; Khaiter, P.A. Application of optimization algorithms for the improvement of water quality monitoring systems. In *Information Technologies in Environmental Engineering*; Springer Berlin: Heidelberg, Germany, 2009; pp. 176–188.
6. Thakur, J.K.; Gossel, W.; Weiß, H.; Wycisk, P. Long-Term Groundwater Monitoring Optimization: Improving Performance and Reducing Costs Associated for a 3d Regionalization of Groundwater Contamination—A Case Study of Bitterfeld/ Wolfen, East Germany. In Proceedings of the Sustainable Water Resource Management and Treatment Technology, Nagpur, India, 19–22 January 2011; Labhasetwar, P., Ed.; NEERI: Nagpur, India, 2011; p. 20.

7. Loaiciga, H.; Charbeneau, R.; Everett, L.; Fogg, G.; Hobbs, B.; Rouhani, S. Review of ground-water quality monitoring network design. *J. Hydraul. Eng.* **1992**, *118*, 11–37.

8. Thakur, J.K.; Gossel, W.; Weiß, H.; Wycisk, P. Optimizing long-term groundwater monitoring network using geostatistical methods. In *Iahs Red Book: Models—Repositories of Knowledge*; IAHS: London, UK, 2012; pp. 133–138.

9. Massmann, J.; Freeze, R.A. Groundwater contamination from waste management sites: The interaction between risk-based engineering design and regulatory policy: 1. Methodology. *Water Resour. Res.* **1987**, *23*, 351–367.

10. Parker, J.; Kim, U.; Kitanidis, P.K.; Cardiff, M.; Liu, X. Stochastic cost optimization of multistrategy dnapl site remediation. *Ground Water Monit. Remediat.* **2010**, *30*, 65–78.

11. MacKenzie, M.C.; Palmer, R.N.; Millard, S.P. Analysis of statistical monitoring network design. *J. Water Resour. Plan. Manag.* **1987**, *113*, 599–615.

12. Ben-Jemaa, F.; Marino, M.A.; Loaiciga, H.A. Multivariate geostatistical design of ground-water monitoring networks.*J. Water Resour. Plan. Manag.* **1994**, *120*, 505–522.

13. Nabi, A.; Gallardo, A.; Ahmed, S. Optimization of a groundwater monitoring network for a sustainable development of the Maheshwaram Catchment, India. *Sustainability* **2011**, *3*, 396–409.

14. Thakur, J.; Diwakar, J.; Singh, S. Hydrogeochemical evaluation of groundwater of Bhaktapur Municipality, Nepal.*Environ. Earth Sci.* **2015**.

15. Datta, B.; Chakrabarty, D.; Dhar, A. Optimal dynamic monitoring network design and identification of unknown groundwater pollution sources. *Water Resour. Manag.* **2009**, *23*, 2031–2049.

16. Chadalavada, S.; Datta, B.; Naidu, R. Uncertainty based optimal monitoring network design for a chlorinated hydrocarbon contaminated site. *Environ. Monit. Assess.* **2011**, *173*, 929–940.

17. Nunes, L.M.; Cunha, M.C.; Ribeiro, L. Groundwater monitoring network optimization with redundancy reduction. *J. Water Resour. Plan. Manag.* **2004**, *130*, 33–43.

18. Dhar, A.; Datta, B. Global optimal design of ground water monitoring network using embedded kriging. *Ground Water***2009**, *47*, 806–815.

19. Meyer, P.D.; Brill, E.D. A method for locating wells in a groundwater monitoring network under conditions of uncertainty. *Water Resour. Res.* **1988**, *24*, 1277–1282.

20. Cieniawski, S.E.; Eheart, J.W.; Ranjithan, S. Using genetic algorithms to solve a multiobjective groundwater monitoring problem. *Water Resour. Res.* **1995**, *31*, 399–409.

21. Reed, P.M.; Minsker, B.S. Striking the balance: Long-term groundwater monitoring design for conflicting objectives. *J. Water Resour. Plan. Manag.* **2004**, *130*, 140–149.

22. Kollat, J.B.; Reed, P.M. Comparing state-of-the-art evolutionary multi-objective algorithms for long-term groundwater monitoring design. *Adv. Water Resour.* **2006**, *29*, 792–807.

23. Stollberg, R. *Groundwater Contaminant Source Zone Identification at an Industrial and Abandoned Mining Site: A Forensic Backward-in-Time Modelling Approach*; Martin Luther University: Halle (Saale), Germany, 2013.

24. Heidrich, S.; Schirmer, M.; Weiss, H.; Wycisk, P.; Grossmann, J.; Kaschl, A. Regionally contaminated aquifers—Toxicological relevance and remediation options (Bitterfeld case study). *Toxicology* **2004**, *205*, 143–155.

25. Stollberg, R.; Gossel, W.; Wycisk, P.; Weiß, H. Source and Pathway Identification of Groundwater Contaminants Using a Backward Modeling Technique. In Proceedings of the 2nd International FEFLOW User Conference, Potsdam, Germany, 14–18 September; 2009.

26. Simeonov, V.; Stratis, J.; Samara, C.; Zachariadis, G.; Voutsa, D.; Anthemidis, A.; Sofoniou, M.; Kouimtzis, T. Assessment of the surface water quality in northern greece. *Water Res.* **2003**, *37*, 4119–4124.

27. Kumar, M.; Ramanathan, A.; Keshari, A.K. Understanding the extent of interactions between groundwater and surface water through major ion chemistry and multivariate statistical techniques. *Hydrol. Processes* **2009**, *23*, 297–310.

28. Wold, S.; Esbensen, K.; Geladi, P. Principal component analysis. *Chemom. Intell. Lab. Syst.* **1987**, *2*, 37–52.

29. World Health Organization (WHO). *Guidelines for Drinking-Water Quality*, 4th ed.; WHO: Geneva, Swizerland, 2011.

30. Environmental Protection Agency (EPA). National Primary Drinking Water Regulations. Available online: http:// water. epa.gov/drink/contaminants/index.cfm (accessed on 25 December 2012).

31. Weiss, H.; Schirmer, M.; Merkel, P. Development of novel reactive barrier technologies at the safira test site, Bitterfeld. In *Advanced Groundwater Remediation: Active and Passive Technologies*; Thomas Telford Publishing: London, UK, 2002; pp. 159–172.

32. Haase, G.; Händel, L.; Nagel, C.; Opp, C.; Zierath, R. Environmental impacts by input of substances in landscapes of the district (Bezirk) of Leipzig, Germany. *GeoJournal* **1990**, *22*, 153–165.

33. Moisey, J.; Fisk, A.T.; Hobson, K.A.; Norstrom, R.J. Hexachlorocyclohexane (hch) isomers and chiral signatures of α-

hch in the arctic marine food web of the northwater polynya. *Environ. Sci. Technol.* **2001**, *35*, 1920–1927.

34. Wycisk, P.; Stollberg, R.; Neumann, C.; Gossel, W.; Weiss, H.; Weber, R. Integrated methodology for assessing the hch groundwater pollution at the multi-source contaminated mega-site bitterfeld/wolfen. *Environ. Sci. Pollut. Res.* **2013**, *20*, 1907–1917.

35. Cameron, K.; Hunter, P.M. *Geostatistical Temporal-Spatial (Gts) 1.0.0.*; Steps, G., Ed.; MacStat Consulting, Ltd.: Colorado, CO, USA, 2010. [

36. Cameron, K.; Hunter, P. Using spatial models and kriging techniques to optimize long-term ground-water monitoring networks: A case study. *Environmetrics* **2002**, *13*, 629–656.

37. Isaaks, E.; Srivastava, R. *Applied Geostatistics*; Oxford University Press: New York, NY, USA, 1989; p. 561.

38. Gilbert, R.O. *Statistical Methods for Environmental Pollution Monitoring*; Wiley: Hoboken, NJ, USA, 1987.

39. Sen, P.K. Estimates of the regression coefficient based on Kendall's tau. *J. Am. Stat. Assoc.* **1968**, *63*, 1379–1389. [

40. Brauner, J.S. Non-Parametric Estimation of Slope: Sen's Method in Environmental Pollution. Environmental sampLing and Monitoring Primer. Available online: http://www.webapps. cee. vt.edu/ewr/environmental/teach/smprimer/sen/sen.html (accessed on 25 June 2012).

41. Mann, H.B. Nonparametric tests against trend. *Econom. J. Econom. Soc.* **1945**, *13*, 245–259. [

42. Kendall, M.G.; Stuart, A. *The Advanced Theory of Statistics: Design and Analysis, and Time-Series*; Macmillan Pub. Co.: London, UK, 1976; Volume 3.

43. Software, G. Surfer 11. Available online: http:// www.goldensoftware.com/products/surfer (accessed on 5 December 2011).

44. Hubert, T. *Methodische Ansätze zur Geologischen 3d-Modellierung Oberflächennaher Schichtenfolgen und Ihre Auswirkung auf die Numerische Grundwassermodellierung*; Martin Luther Univesity: Halle, Germany, 2011. (In German)

45. Thakur, J.K. *Methods in Groundwater Monitoring: Strategies Based on Statistical, Geostatistical, and Hydrogeological Modelling and Visualization*; Martin-Luther University: Halle-Wittenberg, Halle (Saale), Germany, 2013.

46. Thakur, J.K.; Gossel, W.; Weiß, H.; Wycisk, P. Improving groundwater monitoring strategies based on groundwater monitoring network optimization. In *TASK—The Centre of Competence for Soil, Groundwater and Site Revitalisation*; Helmholtz Centre for Environmental Research - UFZ: Liepzig, Germany, 2011.

47. Singh, S.K.; Singh, C.K.; Kumar, K.S.; Gupta, R.; Mukherjee, S. Spatial-temporal monitoring of groundwater using multivariate statistical techniques in Bareilly district of Uttar Pradesh, India. *J. Hydrol. Hydromech.* **2009**, *57*, 45–54.

48. Weiß, H.; Schirmer, M.; Teutsch, G.; Merkel, P. Sanierungsforschung in regional kontaminierten aquiferen (safira)—2. Projektüberblick und pilotanlage. *Grundwasser* **2002**, *7*, 135–139. (In German).

49. Weiß, H.; Teutsch, G.; Fritz, P.; Daus, B.; Dahmke, A.; Grathwohl, P.; Trabitzsch, R.; Feist, B.; Ruske, R.; Böhme, O. Sanierungsforschung in regional kontaminierten aquiferen (safira)—1. Information zum forschungsschwerpunkt am standort bitterfeld. *Grundwasser* **2001**, *6*, 113–122. (In German).

50. Knotters, M.; Brus, D.; Oude Voshaar, J. A comparison of kriging, co-kriging and kriging combined with regression for spatial interpolation of horizon depth with censored observations. *Geoderma* **1995**, *67*, 227–246.

51. Bergström, S.; Graham, L.P. On the scale problem in hydrological modelling. *J. Hydrol.* **1998**, *211*, 253–265.

52. Dooge, J.C. Scale problems in hydrology. *Spec. Publ.* **1997**, *48*, 84–145.

53. Gupta, V.K.; Rodríguez-Iturbe, I.; Wood, E.F. Runoff generation and basin response. In *Scale Problem in Hydrology*; Springer Netherlands: Rotterdam, Netherlands, 1986.

54. Sivapalan, M.; Kalma, J.D. Scale problems in hydrology: Contributions of the robertson workshop. *Hydrol. Process.* **1995**, *9*, 243–250.

55. Sivapalan, M.; Grayson, R.; Woods, R. Scale and scaling in hydrology. *Hydrol. Processes* **2004**, *18*, 1369–1371.

56. Press, W.H.; Flannery, B.P.; Teukolsky, S.A.; Vetterling, W.T. *Numerical Recipes in Fortran 77: Volume 1, Fortran Numerical Recipes: The Art of Scientific Computing*; Cambridge University Press: Cambridge, UK, 1992; Volume 1.

57. Press, W.H.; Teukolsky, S.A.; Vetterling, W.T.; Flannery, B.P. *Numerical Recipes 3rd Edition: The Art of Scientific Computing*; Cambridge University Press: Cambridge, UK, 2007.

58. Storey, R.G.; Howard, K.W.; Williams, D.D. Factors controlling riffle-scale hyporheic exchange flows and their seasonal changes in a gaining stream: A three-dimensional groundwater flow model. *Water Resour. Res.* **2003**.

59. Wroblicky, G.J.; Campana, M.E.; Valett, H.M.; Dahm, C.N. Seasonal variation in surface-subsurface water exchange and lateral

hyporheic area of two stream-aquifer systems. *Water Resour. Res.* **1998**, *34*, 317–328.

60. Thakur, J.K.; Gossel, W.; Weiß, H.; Wycisk, P. Optimizing Long Term Groundwater Monitoring Network in a Complex Contaminated Site. In Proceedings of the HIGRADE Conference, Leipzig, Germany, 26 March 2012; Helmholtz Centre for Environmental Research (UFZ): Leipzig, Germany, 2012.

2

# INTEGRATING EXPERT KNOWLEDGE WITH STATISTICAL ANALYSIS FOR LANDSLIDE SUSCEPTIBILITY ASSESSMENT AT REGIONAL SCALE

**Christos Chalkias [1,*], Christos Polykretis [1,†], Maria Ferentinou [2,†] and Efthimios Karymbalis [1,†]**

[1]Department of Geography, Harokopio University, El. Venizelou 70, Athens 17671, Greece

[2]Department of Geological Sciences, University of KwaZulu-Natal, Westville, Private Bag X54001, South Africa

## ABSTRACT

In this paper, an integration landslide susceptibility model by combining expert-based and bivariate statistical analysis (Landslide Susceptibility Index—LSI) approaches is presented. Factors related with the occurrence of landslides—such as elevation, slope angle, slope aspect, lithology, land cover, Mean Annual Precipitation (MAP) and Peak Ground Acceleration (PGA)—were analyzed within a GIS environment. This integrated model produced a landslide susceptibility map which categorized the study area according to the probability level of landslide occurrence. The accuracy of the final map was evaluated by Receiver Operating Characteristics (ROC) analysis depending on an independent (validation) dataset of landslide events. The prediction ability was found to be 76% revealing that the integration of statistical analysis with human expertise can provide an acceptable landslide susceptibility assessment at regional scale.

## KEYWORDS

integration modeling; landslide susceptibility; expert-based fuzzy weighting; bivariate statistics; GIS; Greece

## 1. INTRODUCTION

Natural hazards have been studied continuously, and systematically, and during the last three decades, their frequency seems to be on the rise [1]. The need for reliable hazard maps is uncontested, in order to mitigate their effects in both natural and manmade environments. From 1900 to 2014, nearly 6% of global natural hazards were associated with landslides [2]. Mountainous or rural environments are subject to landslides as a major natural hazard which causes substantial economic, human, and environmental losses throughout the world each year [3].

Landslide Susceptibility (LS) is defined as the likelihood of a landslide occurring in an area on the basis of local terrain conditions [4]. LS zonation refers to the division of the land in homogeneous areas or domains according to the degree of actual or potential hazard [5]. In the last years, LS mapping has become a valuable tool for assessing disaster phenomena. The cartographic products of LS modeling can be used for developing mitigation plans and selecting the more suitable locations for construction.

Given the powerful abilities of Geographic Information Systems (GIS) technology to deal with a large amount of spatial data, LS is analyzed through effective spatial analysis tools. Many GIS-based approaches have been developed for LS assessment. In general, they are divided into two groups: qualitative and quantitative methods. The most important difference between these two types of methods is their degree of objectivity. The qualitative methods, used extensively during the 1970s and 1980s, depend on the knowledge and previous experience of the experts, resulting in a high degree of subjectivity. They include the geomorphological analysis [6,7] and the use of index or parameter maps. The use of index or parameter maps is subdivided into two methods: the combination or overlay of index maps [8,9] and the logical analytical models [10,11]. Some qualitative approaches, however, incorporate the idea of ranking and weighting the parameters involved, and may be semi-quantitative in nature [12]. Examples of this kind of approach are the Analytical Hierarchy Process (AHP) [13,14], the Weighted Linear Combination (WLC) [15], and their combination [16].

The quantitative methods are based on numerical expressions of the relationship between controlling factors and landslide occurrence. They include statistical analysis, geotechnical engineering approaches [17,18], and soft computing techniques such as Artificial Neural Network (ANN) [19,20,21,22], neuro-fuzzy [23,24], support vector machine [16,23] and decision tree [23,25] methods. In statistical analysis methods, the combination of parameters that have led to landslides in the past are determined statistically, and quantitative predictions are made for areas not affected by landslides but where similar background conditions exist [26]. Several researchers have applied these methods, which involve either bivariate [27,28,29,30] or multivariate analysis [16,31,32,33]. In bivariate analysis, each individual landslide-influencing factor is combined with a landslide inventory map, and weight values based on landslide densities are calculated for its corresponding categories. A well-known and widely used bivariate analysis method for LS zonation is the Landslide Susceptibility Index (LSI) [14,22,34,35]. It is increasingly recognized that in both qualitative and quantitative models, the results are prone to the inherent uncertainties related to various analysis parameters such as the errors and variability in model choice, weighting factors, data used, and human judgment [15].

In recent years, another soft computing technique, known as fuzzy logic, has been applied to map LS. Considering the determination of weightings (fuzzy membership values), this technique is either expert-based (semi-quantitative) approach or data-driven (quantitative) approach or a combination of both [36]. In expert-based approaches

[37,38], the weightings are assigned based on the field knowledge of the experts, whereas in data-driven approaches [39,40] the weightings are determined according to correlations between landslide locations and landslide-conditioning factors. Such an expert-based approach, referred as Trapezoidal Fuzzy Number Weighting (TFNW) [12,41,42], was used in this study.

In Greece, previous studies concerning LS assessment have been implemented on moderate scale (1:25,000–1:500,000) using semi-quantitative methods such as AHP, WLC and TFNW [12,43,44,45], statistical analysis [22,35,46,47], ANN [22,48] and integration methods [49], and on national scale (<1:1,000,000) using statistical analysis [50,51], ANNs [50] and fuzzy logic [52].

The main aim of this paper was to improve our previous research works [12,35], integrating two different and already tested LS assessment methods: LSI and TFNW. Thus, a GIS-based integration method was developed and presented at regional scale (1:500,000) for Peloponnese peninsula, Greece. Furthermore, a validation analysis was implemented to estimate the prediction ability of the proposed method.

## 2. STUDY AREA

Peloponnese was selected as the study area for the proposed method. It is one of the nine geographical departments of Greece and is located in its southern part (Figure 1). Peloponnese constitutes the largest peninsula of the country with an extent of 21,439 km$^2$. It is divided into seven administrative prefectures: Achaia, Argolida, Arkadia, Ilia, Korinthia, Lakonia and Messinia (Figure 1), with a population of 1,086,935 inhabitants [53]. Peloponnese has a typical Mediterranean climate with a hot and relatively dry summer, and a wet season during autumn, winter and spring [54]. The climate as well as the geotectonic status, the lithology and the seismicity have mainly formed the geomorphology of the study area. In general, its geomorphology can be characterized as complex, with a mountainous interior, many coastal cliffs in the south, and basins, coastal beaches, lakes and inland basins on the west and southeast coasts.

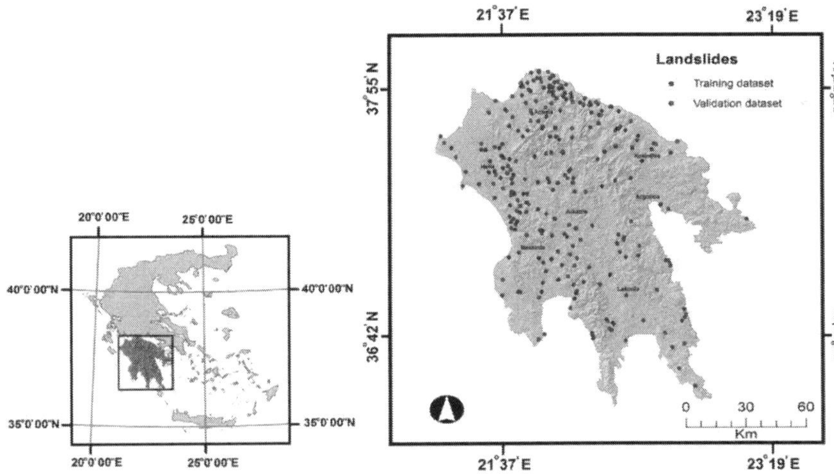


**Figure 1.** The location of the study area (Peloponnese peninsula) and the landslide inventory.

Peloponnese is a region highly vulnerable to the occurrence of severe natural disasters such as earthquakes, floods and forest fires. However, due to its complex geomorphology, all causal and triggering factors of landslide occurrence present a high spatial variability. Considering the morphological and geological conditions of Peloponnese, it is evident that slope stability is one of the most severe and frequent hazards for public safety as well as land-use planning and management of the area.

# 3. MATERIALS AND METHODS

In order to accomplish the LS analyses in the study area, a spatial database was designed and developed, and spatial analysis tools were implemented within GIS environment with the use of an ArcGIS (version 9.3) software package. This database comprises two main parts: (a) the landslide inventory dataset and (b) the datasets for the landslide-influencing factors.

## 3.1. Landslide Inventory

In the present study, two main landslide inventories were used: (a) an inventory maintained by Institute of Geology and Mineral Exploration (IGME) covering a time period from 1910 to 1995 [55] and (b) an inventory, developed on the basis of fieldwork and aerial photography [56] interpretation, which contains landslide events occurring from 1995 to 2003. The final landslide dataset consists of 282 landslides (217 and 65 landslides from the first and second inventory, respectively) presented as point features in the centroid of each pixel (Figure 1). This dataset was randomly split into two separate groups: a training dataset and a validation dataset. The training dataset (75% of the landslide inventory) was used for the implementation of the model, whereas the validation dataset (25% of the landslide inventory) was used during the verification of the results produced from the model (see among others [57,58,59]). Our intention was to keep the same sampling strategy (*i.e.*, the same percentages for training and validation datasets) for both the integration model and the single models [12,35] combined for its creation in order to increase their comparability.

The landslides under investigation have more or less the same characteristics: lateral based and downslope movement of soils or rocks. In line with [60], in this study, the term landslide is used for translational and rotational earth slides, which were recorded in landslide inventories. These events vary consistently in volume, from some thousands of $m^3$ to several million $m^3$ [51], and depicting small to extremely large magnitude according to [61] classification. Subaqueous and liquefaction events are not examined.

## 3.2. Landslide-Influencing Factors

Although there are no standard guidelines for selecting the landslide-influencing factors, the nature of the study area, the scale of the analysis, the data availability and the general literature guidelines were taken into account [26,62,63,64,65]. In this study, land cover, lithology, elevation, slope angle, slope aspect, MAP and PGA have been selected as the landslide-influencing factors.

Slope stability is highly influenced by land cover. The land cover layer was classified by following the level-1 classification scheme of the CORINE Coordinate of Information on the Environment) program data [66]. Agricultural, forest and semi-natural areas cover the main part of Peloponnese, whereas urban is the dominant land cover in the coastal zone of the peninsula.

Lithology plays a key role in landslide activity since different lithologic units have different slope stability performances [67]. The lithological layer of the study area was created from the Geological Map of Greece [68]. The main lithological formations are carbonate rocks (44%): limestones, dolomites and marbles, and Neogene sediments (22%): usually marls, sandstones and mudstones. The central and western part of Peloponese belongs to the External Hellenides, which is characterized by Mesozoic–early Cenozoic sedimentary units (mainly carbonate rocks, schists-cherts and flysch) [69]. The most critical landslide-prone formations are flysch and neogene sediments, while schists and cherts significantly contribute in landslide phenomena [51].

Among all parameters for LS zonation, elevation, slope angle and slope aspect have been recognized as the most important [70,71,72]. The elevation dataset is useful to classify the local relief and locate points of maximum and minimum heights within terrains. Furthermore, it is well supported in the literature [43,46] that slope gradients have a large impact on landsliding in Peloponnese. In the study area, the slopes vary from gentle to very steep. Most of the landslides occur on gentle natural slopes where the translational type predominates. The aspect parameter is also related to differential weathering, exposure to sunlight and drying winds and soil moisture [62]. The Digital Elevation Model (DEM) was the key to generating the topographic factors related to the landslide activity of the area under investigation. We used the DEM from SRTM (Shuttle Radar Topography Mission) database. Elevation (0–2367 m), slope angle (0°–54°) and slope aspect were extracted from this DEM.

Previous studies emphasized the need of incorporating "dynamic" factors (MAP, PGA) with other "static" factors for LS zonation mapping in areas, whereas these factors are playing an important role not only in the reactivation of old landslides but in the development of new ones [73,74]. It is, therefore, imperative to incorporate these parameters, while carrying out LS analysis in seismically active and frequently rainfall-influenced regions. The precipitation data (339–1655 mm) used in this study refers to the mean annual precipitation during the period from 1950 to 1974 (source: Public Power Company, Athens, Greece). MAP is the average of the available long-term records [75]. The study area belongs to a seismically active zone with tectonism expressed through faults, thrust zones and folds. The seismic factor (0.05–3.95 $m/s^2$) was produced from the map of expected peak ground acceleration with a 475-year return period (10% probability of exceedance in 50 years). This map, from which the corresponding layer was derived, was launched in 1992 from the Technical Chamber of Greece. PGA is the absolute maximum amplitude of recorded ground acceleration [76]. Heavy rainfall and earthquakes have triggered several landslide events, mainly in northern and western areas of the peninsula [43,77,78]. Rainfall triggered landslides usually are rapid short moving events, while slow short-moving type also occur including extensive instability zones [79]. On the other hand, seismic triggered landslides occur in the vicinity of active faults, and usually related to other secondary seismic events, like soil liquefaction, subsidence of the coastal strip, and rock falls [80,81].

Most of the final layers were in raster format (grid), while others were converted from vector (point, line or polygon) to raster format (with cell size 250 m × 250 m for land cover, lithology, MAP and PGA factors, and 90 m × 90 m for elevation, slope angle and slope aspect factors).

## 3.3. Methodology

A GIS-based integration approach was developed to create a LS map using two different methods: LSI and TFNW. Such integration can adjust the parameters included in this study by means of the precise LS assessment and effectively reduce the subjectivity and uncertainty resulting from the use of single method. The LSI method was proposed by [82] and is based on the relationship between the spatial distribution of landslides and each conditioning factor [62]. Here, in order to quantify the impact of various categories for the standardization of each factor, we used the TFNW. The details of this approach have been described by [41]. The main steps of this methodology are presented below:

(a)Categorization of all landslide-influencing factors. In this step, the "Natural Breaks (Jenks)" categorization (five categories) was implemented for the "dynamic" factors (MAP and PGA), whereas for the other two factors with continuous values (elevation and slope angle) the categorization was executed in a manual way based on their presented values. "Natural Breaks (Jenks)" method is a data classification method that minimizes variance within groups of data and maximizes variance between groups of data [83]. For land cover, lithology, and aspect, all the categories of the nominal scale were preserved.

(b)Calculation of LSI for all factor categories (Table 1) based on Equation (1) [26]:

$$LSI_{i,j} = ln\left(\frac{N_{i,j}}{A_{i,j}} / \frac{N_T}{A_T}\right) \quad (1)$$

with $N_{i,j}$ the number of landslides in category j of the factor i, $A_{i,j}$ the area of this category, $N_T$ the total number of landslides and $A_T$ the total area under investigation. Thus, the calculation of the landslide density in each category and of the area for each category was implemented using GIS-based overlay functions.

**Table 1.** Categories, landslide densities, susceptibilities (LSI), fuzzy values and weights for landslide related factors.

| Layers (Factors) | Categories (Classes) | Landslide Density | LSI | (Experts), STAT Fuzzy Value | LSI/TFNW Weight |
|---|---|---|---|---|---|
| Land cover | Artificial surfaces | 0.09 | 2.04 | (H, M, M), VH | 0.45 |
| | Agricultural areas | 0.66 | 0.42 | (M, H, H), H | 0.42 |
| | Forest and semi-natural land | 0.25 | -0.79 | (L, L, M), VL | 0.13 |
| Lithology | Phyllites/Gneiss (metamorphic) | 0.07 | 0.16 | (L, M, L), H | 0.14 |
| | Limestones—Marbles | 0.20 | -0.77 | (L, L, L), VL | 0.05 |
| | Schists (metamorphic) | 0.01 | -0.51 | (M, M, M), VL | 0.08 |
| | Neogene | 0.44 | 0.69 | (H, VH, M), VH | 0.22 |
| | Quaternary | 0.12 | -0.06 | (H, VH, M), M | 0.16 |
| | Flysch | 0.09 | 0.06 | (VH, VH, VH), M | 0.19 |
| | Cherts—Schists | 0.06 | 0.36 | (M, L, M), H | 0.15 |
| MAP | <663 mm | 0.03 | -1.69 | (L, VL, M), VL | 0.07 |
| | 663–884 mm | 0.30 | 0.14 | (M, L, H), M | 0.19 |
| | 885–1079 mm | 0.43 | 0.29 | (M, M, H), H | 0.25 |
| | 1080–1295 mm | 0.19 | -0.03 | (H, H, VH), M | 0.25 |
| | >1295 mm | 0.05 | -0.22 | (VH, VH, VH), L | 0.23 |
| PGA | <1.16 $m/s^2$ | 0.02 | -0.14 | (L, VL, L), M | 0.16 |
| | 1.16–1.88 $m/s^2$ | 0.01 | -0.98 | (L, L, L), VL | 0.08 |
| | 1.89–2.53 $m/s^2$ | 0.17 | -0.81 | (M, M, M), VL | 0.13 |
| | 2.54–3.11 $m/s^2$ | 0.67 | 0.53 | (H, H, H), VH | 0.37 |
| | >3.11 $m/s^2$ | 0.13 | -0.28 | (VH, VH, VH), L | 0.27 |
| Elevation | <234 m | 0.36 | 0.17 | (L, VL, L), H | 0.25 |
| | 234–524 m | 0.27 | 0.11 | (L, L, M), M | 0.22 |
| | 525–851 m | 0.22 | -0.03 | (H, M, M), M | 0.28 |
| | >851 m | 0.15 | -0.43 | (M, H, H), L | 0.24 |
| Slope angle | <5° | 0.31 | -0.16 | (VL, VL, L), L | 0.09 |
| | 5°–10° | 0.30 | 0.19 | (L, L, L), H | 0.21 |
| | 11°–15° | 0.22 | 0.23 | (M, M, M), H | 0.25 |
| | 16°–20° | 0.12 | 0.08 | (H, H, H), M | 0.25 |
| | >20° | 0.05 | -0.77 | (VH, VH, VH), VL | 0.21 |
| Slope aspect | North | 0.31 | 0.31 | (H, H, M), H | 0.31 |
| | East | 0.21 | -0.15 | (M, M, M), M | 0.23 |
| | South | 0.16 | -0.47 | (M, M, M), L | 0.18 |
| | West | 0.32 | 0.16 | (H, H, H), M | 0.28 |

Table 1. *Cont.*

LSI: Landslide Susceptibility Index, LSI/TFNW: The integration method, VL: Very Low, L: Low, M: Moderate, H: High, VH: Very High.

(c)The maximum value, the range and the standard deviation of LSI values for all factors were calculated in order to interpret the importance of each factor. The range is the difference between the maximum and minimum LSI values. The standard deviation measures the spread of the LSI values about the mean value.

(d)Definition of linguistic variables and fuzzy numbers for LS categories in order to incorporate uncertainty in the analysis. All fuzzy numbers were expressed as ($a_k$, $b_k$, $c_k$, $d_k$). The definition of these fuzzy numbers is presented in [12].

(e)Three independent experts, with scientific background in the fields of geosciences and engineering geology, and with experience and scientific knowledge of the study area, were invited to assign a linguistic importance weighting for every category of each factor. From these linguistic judgments we obtained the corresponding fuzzy numbers. Such judgments are inevitably subjective, but, by proposing several possible scenarios, followed by the systematic testing and elimination of options, as a result of additional investigation and discussion, it is possible to develop reliable estimates. Experimental evidence suggests that group judgments appear to be more accurate than judgments of a typical group member [84]. In this case, we considered a homogeneous group of experts with equal degree of importance for each one. Accordingly, the overall expert-based judgment is the mean value of the three judgment values. The sum of these numbers is still a fuzzy number. Thus, we proceeded to the computation of the aggregated fuzzy weights of individual categories (Table 1).

(f)A fourth "statistical expert" judgment was added in the above list on the basis of the statistical analysis (Table 1). This statistical interpretation was based on the LSI values for each category according to the following rules:

- In a category with LSI value between –0.15 and 0.15, a "Moderate" susceptibility was assigned.
- Categories with negative (<–0.15) LSI values were defined as zones of "Very Low" or "Low" susceptibility according to their ascending order.
- Categories with positive (>0.15) LSI values were defined as zones of "Very High" or "High" susceptibility according to their descending order.

Accordingly, the range of LSI values was interpreted for each factor according to the following rules:

- In a factor with range between 1.00 and 1.50, a "Moderate" susceptibility was assigned.
- Factors with range <1 were characterized with "Very Low" or "Low" susceptibility according to their ascending order.
- Factors with range >1.50 were characterized with "Very High" or "High" susceptibility according to their descending order.

It is worth mentioning that in our case, we assigned an equal degree of importance between the "statistical expert" and the group of experts.

(g)After the defuzzification of the fuzzy weights of individual LS categories, we proceed to the computation of the normalized weights and the construction of the weight vector (Table 1). More details of this step could be found in [12].

By implementing the same procedure described previously, we used expert-based and statistical-based linguistic variables to calculate the weight of importance for each conditioning factor (Table 2).

(h)The last step is the aggregation of relative values, and the generation of the final LS map (Figure 2). This step was implemented by using the WLC method [85]. We classified the final LS map into five discrete categories: "Very Low", "Low", "Moderate", "High" and "Very High" landslide susceptibility according to the standard deviation classification method. This method uses the mean value to generate class breaks by adding or subtracting one standard deviation at a time [70]. Moreover, in order to maintain five categories, we embedded low and high outliers into "Very Low" and "Very High" susceptibility categories, respectively.

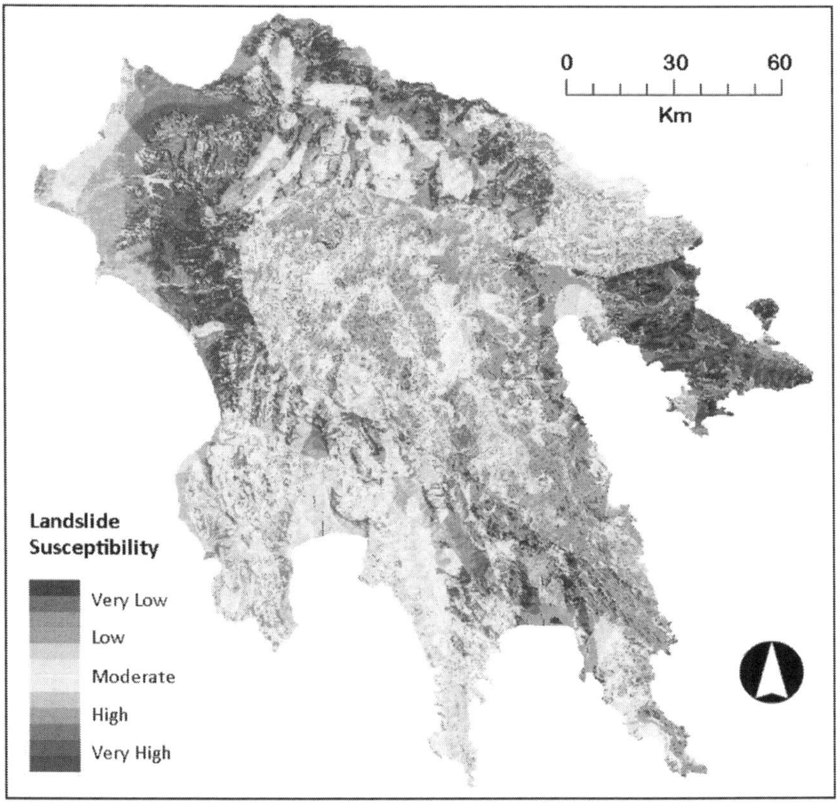


**Figure 2.** The landslide susceptibility map produced from the integration LSI/TFNW model.

## 4. RESULTS—VALIDATION

As mentioned previously, we developed an integration LS model according to the LSI/TFNW approach. This approach allowed for the assessment of landslide susceptibility by integrating expert-based and statistical modelling methods.

The results from the LSI/TFNW model are presented in Table 1 and Table 2. The most important causal factors were found to be MAP, slope angle and lithology with weight values 0.20, 0.18 and 0.17, respectively.

Elevation was the weakest factor in the scale of importance (weight value = 0.08).

**Table 2.** Weight values for each factor according to the integration LSI/TFNW method.

| Parameter | Parameter Importance (Expert Judgment), STAT | LSI/TFNW Weight |
|---|---|---|
| Land cover | (M, M, H), M | 0.12 |
| Lithology | (VH, VH, H), H | 0.17 |
| MAP | (VH, VH, VH), VH | 0.20 |
| PGA | (H, H, H), H | 0.16 |
| Elevation | (M, L, H), L | 0.08 |
| Slope angle | (VH, VH, VH), H | 0.18 |
| Slope aspect | (H, M, M), L | 0.09 |

The output LS map (Figure 2) from the integration model shows that 26% and 7% of the study area were classified as "High" and "Very High" susceptibility zones, respectively. The same map also shows that the zones of "High" and "Very High" susceptibility are mainly located around the northern (coastal) and western part of Peloponnese, with pockets of "High" susceptibility in the rest part of the peninsula.

The overlay of the final LS map with the landslide training dataset indicated that 28%, 42% and 24% (total: 94%) of the landslide events fall within "Very High", "High" and "Moderate" LS zones (in total 66% of the study area), respectively. It is notable that according to the used model only 6% of the dataset falls in "Low" susceptibility zone, with no landslides into the "Very Low" susceptibly zone.

We have also proceeded with a standard validation analysis using data from landslide events, not included in the initial spatial database (validation dataset), in order to estimate the overall performance of the model in the study area. Validation is essential to know the predictive value of the model [86]. For the validation of the output, the Receiver Operating Characteristics (ROC) curve was drawn, and the Area Under

Curve (AUC) value was calculated for the proposed model. ROC analysis is considered as a powerful method for the validation of LS models [87,88]. In practice, the AUC performs very well and is often used when a general measure of predictiveness is desired [89]. The AUC values range from 0.5 to 1.0. The ideal model yields a value close to 1.0 (perfect fit), whereas a value close to 0.5 indicates an inaccurate model (random fit).

Figure 3 shows the ROC curve of LSI/TFNW model for the validation dataset. The AUC value of 0.76 indicates a reasonable prediction ability of the model.

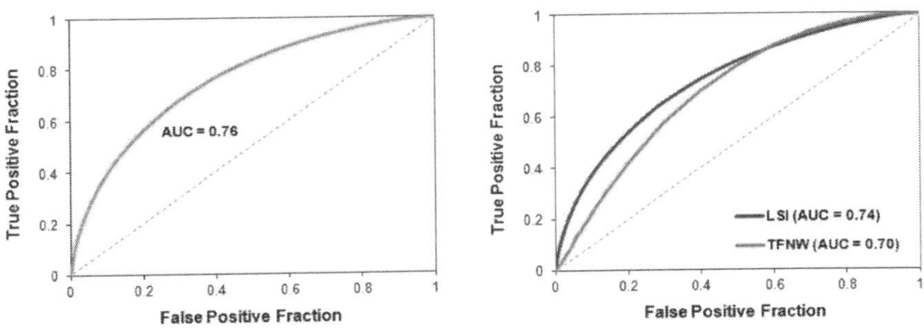


**Figure 3.** ROC curves for the integration LSI/TFNW model and the combined single models, LSI and TFNW.

Previous studies [12,35] in the same area with the use of the single methods combined in this integration model (TFNW and LSI, respectively), produced slightly worse results (Figure 3). The AUC values for these models were 0.70 and 0.74, respectively, based on the standard deviation classification method for their final LS maps.

Specifically, comparing the final map of the proposed integrated method with the landslide susceptibility map produced from TFNW model [12], it is shown that in both maps 14% (3039 $km^2$), 9% (2001 $km^2$) and 1% (301 $km^2$) of the study area are defined as zones of "Moderate", "High" and "Very High" susceptibility, respectively. On the other hand, the differences between the LS maps for these three susceptibility zones are presented to cover 32% (6773 $km^2$) of the study area. The differences for the remaining susceptibility zones ("Very Low" and "Low") cover a restricted part of the study area (7% or 1377 $km^2$).

In addition, comparing the final LS map of this study with this one created from LSI model [35], it is derived that in both maps 21% (4427 $km^2$), 18% (3829 $km^2$) and 4% (881 $km^2$) of the study area belong to "Moderate", "High" and "Very High", respectively, susceptibility zones. On the contrary, the coverage of differences between the final maps for these three susceptibility zones is found to be 16% (3291 $km^2$) of the study area. Finally, the coverage of differences for the remaining susceptibility zones is more or less similar with the relative coverage of the aforementioned comparison between integration and TFNW models.

# 5. DISCUSSION

The idea of implementing an integration method for LS zonation is not new. Recent examples are the efforts of [90,91,92]. In the first study, AHP and fuzzy sets were combined to determine LS levels for areas of the Metro Vancouver region, British Columbia, Canada. The parameters were standardized to a common measurement scale using fuzzy set membership functions and the weight value for each parameter was calculated using AHP. In the second study, the authors produced a LS zonation map in the Chehelchay Basin (northeastern Iran). They applied Geographically-Weighted Principal Component Analysis (GWPCA) to calculate LS using a fuzzy gamma operator. Finally, the third study focused on the methodology of the urgent LS assessment right after the 2013 Lushan earthquake (southwest of Sichuan Province, China). An assessment approach was developed integrating bivariate statistical analysis (LSI) and AHP methods for the assignment of factor class weights and factor weights, respectively.

This study combined an expert-based semi-quantitative approach (TFNW) with a data-driven statistical method (LSI) to prepare a LS map at a regional scale on the Peloponnese peninsula, Greece. The main motivation is to create an integration model that preserves and integrates the advantages of the combined models at the specific scale of analysis. To achieve this objective, seven conditioning factors (elevation, slope angle, slope aspect, lithology, land cover, MAP and PGA) were taken into consideration.

The LSI model is a data-driven bivariate statistical approach. The bivariate statistical technique is one of the most preferable models at this scale [93]. In this LS modeling, all the problems of landslide inventory maps will automatically be inherited to the final susceptibility maps [94]. In general, a reliable, accurate and complete landslide inventory map will provide high-quality statistical models [95]. In many countries—including Greece—this kind of extended landslide inventory is not available.

On the other hand, for the expert-based fuzzy weighting model, the landslide inventory map is not needed. This kind of analysis is purely subjective. To some extent, opinions may change for every individual expert and thus may be subjected to cognitive limitations with uncertainty and subjectivity. However, methods that depend on expert opinions are often useful for regional assessments [96]. A significant part of the semi-quantitative LS research follows a similar strategy as the proposed one by inviting a limited number of experts (see among others [42,97]). The main issue is not to invite as many scientists as possible but to invite experts with detailed knowledge of the landslide events in the area under investigation and to ensure an overall consensus of their evaluation about the importance of the factors involved.

Recently, ROC analysis has been widely used in the international literature to evaluate the prediction capability of the LS mapping methods. In this study, the empiric ROC area for the integration model was estimated to be 0.76 for the validation dataset. Thus, there is 76% agreement between prepared LS map and landslide locations of the validation dataset, which is a reasonable result, taking into consideration the scale of analysis. Moreover, this model proved to be slightly more effective for the overall LS assessment of the study area as it recorded the highest percentage in our validation test in comparison with LSI and TFNW methods (74% and 70%, respectively). Therefore, the reliability of the proposed integration model is directly related to the reliability of both data-driven and expert-based methods combined. In other words, it seems that a combination of two quite satisfactory methods according to the proposed integration model generates even better results. However, this preliminary finding needs more research to be confirmed through the implementation of the proposed model in more studies areas with various geographical and geotectonic settings.

The implementation of the integration model in the study area revealed that there are different zones within Peloponnese which seem to configure various landslide susceptibility clusters. The "High" susceptibility areas are concentrated in the western Peloponnese with few scattered pockets in the remaining parts of the peninsula. Additionally, a linear concentration of "High" susceptibility values is located in the northern coastal zone of the study area. According to the final LS map, the "Very High" susceptibility zone covers a significant part of the study area (7% of the total area).

The comparison of the proposed method with the expert-based and the statistical methods showed that the integrated method illustrates more emphatically the zones of "High" or "Low" susceptibility in both of the individual methods. Accordingly, the main part the areas of "High" and "Very High" susceptibility in the hybrid map (classes "4" and "5"), is within the class "5" ("Very high" susceptibility) in the statistical and the expert-based LS maps (99% and 67%, respectively). Concerning the spatial distribution of the areas with high LS values in the integrated method, despite the preservation of the general spatial pattern (high LS values in the northern and the western part of Peloponnese) it has to be noticed that there are some pockets of high LS in areas with limited existence of previous landslides. This fact is due to the expert-based classification and it may be useful to identify high LS zones in areas with lack of landslide events.

The most important factors for the zonation of LS in the study area are MAP, slope angle and lithology. It seems that the incorporation of "dynamic" factors (MAP, PGA) in this regional analysis was more or less beneficial to the LS assessment. Among the categories of these two factors, the categories with high levels of annual precipitation (885–

1079 mm) and seismic acceleration (2.54–3.11 $m/s^2$) have the highest LSI values combining the highest landslide density and coverage area. In contrast with aforementioned categories of MAP and PGA factors, the category with 11°–15° slope angle presents the highest LSI value combining a lower landslide density and coverage area than the categories with low levels of slope angle (<10°).

Some limitations and assumptions of the proposed method have to be pointed out. First of all, as the analysis is based on medium-scale datasets, the results are unsuitable for detailed site-oriented specific analysis. At large scales, more exhaustive datasets and detailed geotechnical information are required. A second limitation is related to the landslide inventory dataset. At regional scale, this dataset does not include the total amount of the landslide events within the area under investigation. In this paper, we considered equal importance between the statistical and expert-based models for the construction of the integration model. Although this is a reasonable starting point, alternative importance ratios should be further tested. Finally, another limitation of this study is that the proposed method does not take into account the mutual relationships between causal factors. Future work will focus on this issue by combining expert-based with multivariate statistical models. Furthermore, it should be mentioned that, according to our analysis, the output LS map presents only the predicted spatial distribution of landslides. It does not present the temporal probability of landsliding.

Therefore, the results from this paper should be used in the first stage of hazard mapping, which is crucial in an essential part of quantitative risk analysis [98,99]. Despite its limitations, the proposed method can produce reliable landslide susceptibility maps at regional scale. This is very useful information for local authorities and decision makers in order to choose suitable locations for future land-use planning and implementation of developments, and target their mitigation strategies.

# 6. CONCLUSIONS

Statistical analysis and expert-based methods are often used to evaluate LS zonation. Considering that these general approaches have their pros and cons when applied at regional scale, we investigate the performance of an integration model. The results obtained show that it is possible to produce satisfactory LS maps with the proposed hybrid model. It seems that the overall assessment provided by the experts efficiently incorporates the knowledge of the study area and of the factors involved in the LS zonation. Thus, to some extent, the expertise of specialists can contribute to the enhancement of the standard statistical method especially in moderate and small scale analyses in areas with incomplete landslide event archives.

Further research may include the validation of the proposed model in other areas characterized by different geomorphological characteristics, and its comparison with other advanced LS models (such as logistic regression and ANN models) at regional scale.

# ACKNOWLEDGMENTS

Authors thank the three experts for their contribution in this paper. Also, authors are grateful to the anonymous reviewers for their useful comments and constructive manuscript reviewing.

# AUTHOR CONTRIBUTIONS

Christos Chalkias conceived of the study, designed the analysis, and participated in its implementation. Christos Polykretis participated in the spatial database creation and the implementation of statistical analysis, and drafted the manuscript. Maria Ferentinou and Efthimios Karymbalis participated in the design and implementation of the study and in the preparation of the validation dataset. Christos Chalkias, Christos Polykretis, Maria Ferentinou and Efthimios Karymbalis participated in the finalization of the manuscript. All authors read and approved the final manuscript.

## CONFLICTS OF INTEREST

The authors declare no conflict of interest.

## REFERENCES


1. Smith, K.; Petley, D.N. *Environmental Hazards: Assessing risk and reducing disaster*, 5th ed.; Routledge: Abingdon, UK, 2009.
2. Centre for Research on the Epidemiology of Disasters (CRED). *EM-DAT: International Disaster Database*; Université Catholique de Louvain: Brussels, Belgium, 2014.
3. Aleotti, P.; Chowdhury, R. Landslide hazard assessment: Summary review and new perspectives. *Bull. Eng. Geol. Environ.* **1999**, *58*, 21–44.
4. Brabb, E.E. Innovative approaches to landslide hazard mapping. In Proceedings of the 4th International Symposium on Landslides, Toronto, ON, Canada, 16–21 September 1984; Volume 1, pp. 307–324.
5. Varnes, D.J. *Landslide Hazard Zonation: A Review of Principles and Practice*; UNESCO: Paris, France, 1984; Volume 3.
6. Santangelo, M.; Cardinali, M.; Rossi, M.; Mondini, A.C.; Guzzetti, F. Remote landslide mapping using a laser rangefinder binocular and GPS. *Nat. Hazards Earth Syst. Sci.* **2010**, *10*, 2539–2546.

7. Listo, F.L.R.; Vieira, B.C. Mapping of risk and susceptibility of shallow-landslide in the city of Sao Paulo, Brazil.*Geomorphology* **2012**, *169–170*, 30–44.

8. Castellanos Abella, E.A.; van Westen, C.J. Qualitative landslide susceptibility assessment bymulticriteria analysis: A case study from San Antonio del Sur, Guantánamo, Cuba. *Geomorphology* **2008**, *94*, 453–466.

9. Avtar, R.; Singh, C.K.; Singh, G.; Verma, R.L.; Mukherjee, S.; Sawada, H. Landslide susceptibility zonation study using remote sensing and GIS technology in the Ken-Betwa River Link area, India. *Bull. Eng. Geol. Environ.* **2011**, *70*, 595–606.

10. Intarawichian, N; Dasananda, S. Analytical hierarchy process for landslide susceptibility mapping in lower Mae Chaem watershed, northern Thailand. *Suranaree J. Sci. Technol.* **2010**, *17*, 277–292.

11. Demir, G.; Aytekin, M.; Akgün, A.; Ikizler, S.B.; Tatar, O. A comparison of landslide susceptibility mapping of the eastern part of the North Anatolian Fault Zone (Turkey) by likelihood-frequency ratio and analytic hierarchy process methods. *Nat. Hazards* **2013**, *65*, 1481–1506.

12. Chalkias, C.; Ferentinou, M.; Polykretis, C. GIS supported landslide susceptibility modeling at regional scale: An expert-based fuzzy weighting method. *ISPRS Int. J. Geo-Inf.* **2014**, *3*, 523–539.

13. Kayastha, P.; Dhital, M.R.; De Smedt, F. Application of the Analytical Hierarchy Process (AHP) for landslide susceptibility mapping: A case study from the Tinau watershed, west Nepal. *Comput. Geosci.* **2013**, *52*, 398–408.

14. Pourghasemi, H.R.; Moradi, H.R.; Fatemi Aghda, S.M. Landslide susceptibility mapping by binary logistic regression, analytical

hierarchy process, and statistical index models and assessment of their performances. *Nat. Hazards* **2013**, *69*, 749–779.

15. Feizizadeh, B.; Blaschke, T. An uncertainty and sensitivity analysis approach for GIS-based multicriteria landslide susceptibility mapping. *Int. J. Geogr. Inf. Sci.* **2014**, *28*, 610–638.
16. Kavzoglu, T.; Sahin, E.K.; Colkesen, I. Landslide susceptibility mapping using GIS-based multi-criteria decision analysis, support vector machines, and logistic regression. *Landslides* **2014**, *11*, 425–439.
17. Michel, G.P.; Kobiyama, M.; Goerl, R.F. Comparative analysis of SHALSTAB and SINMAP for landslide susceptibility mapping in the Cunha River basin, southern Brazil. *J. Soils Sediments* **2014**, *14*, 1266–1277.
18. Dahal, R.K.; Bhandary, N.P.; Hasegawa, S.; Yatabe, R. Topo-stress based probabilistic model for shallow landslide susceptibility zonation in the Nepal Himalaya. *Environ. Earth Sci.* **2014**, *71*, 3879–3892.
19. Wu, X.; Niu, R.; Ren, F.; Peng, L. Landslide susceptibility mapping using rough sets and backpropagation neural networks in the Three Gorges, China. *Environ. Earth Sci.* **2013**, *70*, 1307–1318.
20. Bi, R.; Schleier, M.; Rohn, J.; Ehret, D.; Xiang, W. Landslide susceptibility analysis based on ArcGIS and Artificial Neural Network for a large catchment in Three Gorges region, China. *Environ. Earth Sci.* **2014**, *72*, 1925–1938.
21. Conforti, M.; Pascale, S.; Robustelli, G.; Sdao, F. Evaluation of prediction capability of the artificial neural networks for mapping landslide susceptibility in the Turbolo River catchment (northern Calabria, Italy). *Catena* **2014**, *113*, 236–250.
22. Polykretis, C.; Ferentinou, M.; Chalkias, C. A comparative study of landslide susceptibility mapping using landslide susceptibility

index and artificial neural networks in the Krios River and Krathis River catchments (northern Peloponnesus, Greece). *Bull. Eng. Geol. Environ.* **2015**, *74*, 27–45.

23. Pradhan, B. A comparative study on the predictive ability of the decision tree, support vector machine and neuro-fuzzy models in landslide susceptibility mapping using GIS. *Comput. Geosci.* **2013**, *51*, 350–365.

24. Sdao, F.; Lioi, D.S.; Pascale, S.; Caniani, D.; Mancini, I.M. Landslide susceptibility assessment by using a neuro-fuzzy model: A case study in the Rupestrian heritage rich area of Matera. *Nat. Hazards Earth Syst. Sci.* **2013**, *13*, 395–407.

25. Park, I.; Lee, S. Spatial prediction of landslide susceptibility using a decision tree approach: A case study of the Pyeongchang area, Korea. *Int. J. Remote Sens.* **2014**, *35*, 6089–6112.

26. He, Y.; Beighley, R.E. GIS-based regional landslide susceptibility mapping: A case study in southern California. *Earth Surf. Process. Landf.* **2008**, *33*, 380–393.

27. Pourghasemi, H.R.; Pradhan, B.; Gokceoglu, C.; Mohammadi, M.; Moradi, H.R. Application of weights-of-evidence and certainty factor models and their comparison in landslide susceptibility mapping at Haraz watershed. *Iran. Arab. J. Geosci.* **2013**, *6*, 2351–2365.

28. Epifânio, B.; Zêzere, J.L.; Neves, M. Susceptibility assessment to different types of landslides in the coastal cliffs of Lourinhã (Central Portugal). *J. Sea Res.* **2014**, *93*, 150–159.

29. Sujatha, E.R.; Kumaravel, P.; Rajamanickam, G.V. Assessing landslide susceptibility using Bayesian probability-based weight of evidence model. *Bull. Eng. Geol. Environ.* **2014**, *73*, 147–161.

30. Youssef, A.M.; Al-Kathery, M.; Pradhan, B. Landslide susceptibility mapping at Al-Hasher Area, Jizan (Saudi Arabia)

using GIS-based frequency ratio and index of entropy models. *Geosci. J.* **2015**, *19*, 113–134.

31. Bathrellos, G.D.; Kalivas, D.P.; Skilodimou, H.D. GIS-based landslide susceptibility mapping models applied to natural and urban planning in Trikala, Central Greece. *Estud. Geol.* **2009**, *65*, 49–65.

32. Bai, S.B.; Wang, J.; Thiebes, B.; Cheng, C.; Chang, Z.Y. Susceptibility assessments of the Wenchuan earthquake-triggered landslides in Longnan using logistic regression. *Environ. Earth Sci.* **2014**, *71*, 731–743.

33. Reichenbach, P.; Busca, C.; Mondini, A.C.; Rossi, M. The influence of land use change on landslide susceptibility zonation: The Briga catchment test site (Messina, Italy). *Environ. Manag.* **2014**, *54*, 1372–1384.

34. Bourenane, H.; Bouhadad, Y.; Guettouche, M.S.; Braham, M. GIS-based landslide susceptibility zonation using bivariate statistical and expert approaches in the city of Constantine (Northeast Algeria). *Bull. Eng. Geol. Environ.* **2015**,*74*, 337–355.

35. Chalkias, C.; Ferentinou, M.; Polykretis, C. GIS-based landslide susceptibility mapping on the Peloponnese Peninsula, Greece. *Geosciences* **2014**, *4*, 176–190.

36. Tien Bui, D.; Pradhan, B.; Lofman, O.; Revhaug, I.; Dick, O.B. Spatial prediction of landslide hazards in Hoa Binh province (Vietnam): A comparative assessment of the efficacy of evidential belief functions and fuzzy logic models.*Catena* **2012**, *96*, 28–40.

37. Sujatha, E.R.; Rajamanickam, V. Landslide susceptibility mapping of Tevankarai Ar sub-watershed, Kodaikkanal taluk, India, using weighted similar choice fuzzy model. *Nat. Hazards* **2011**, *59*, 401–425.

38. Zhu, A.X.; Wang, R.; Qiao, J.; Qin, C.Z.; Chen, Y.; Liu, J.; Du, F.; Lin, Y.; Zhu, T. An expert knowledge-based approach to landslide susceptibility mapping using GIS and fuzzy logic. *Geomorphology* **2014**, *214*, 128–138.

39. Pourghasemi, H.R.; Pradhan, B.; Gokceoglu, C. Application of fuzzy logic and Analytical Hierarchy Process (AHP) to landslide susceptibility mapping at Haraz watershed, Iran. *Nat. Hazards* **2012**, *63*, 965–996.

40. Kayastha, P.; Bijukchhen, S.M.; Dhital, M.R.; De Smedt, F. GIS based landslide susceptibility mapping using a fuzzy logic approach: A case study from Ghurmi-Dhad Khola area, eastern Nepal. *J. Geol. Soc. India* **2013**, *82*, 249–261.

41. Wang, W.D.; Xie, C.M.; Du, X.G. Landslides susceptibility mapping based on geographical information system, GuiZhou, south-west China. *Environ. Geol.* **2009**, *58*, 33–43.

42. Wang, W.D.; Guo, J.; Fang, L.G.; Chang, X.S. A subjective and objective integrated weighting method for landslides susceptibility mapping based on GIS. *Environ. Earth Sci.* **2012**, *65*, 1705–1714.

43. Rozos, D.; Bathrellos, G.D.; Skillodimou, H.D. Comparison of the implementation of rock engineering system and analytic hierarchy process methods, upon landslide susceptibility mapping, using GIS: A case study from the Eastern Achaia County of Peloponnesus, Greece. *Environ. Earth Sci.* **2011**, *63*, 49–63.

44. Kritikos, T.; Davies, T.R.H. GIS-based multi-criteria decision analysis for landslide susceptibility mapping at northern Evia, Greece. *Z. Deutsch. Ges. Geowiss.* **2011**, *162*, 421–434.

45. Kouli, M.; Loupasakis, C.; Soupios, P.; Rozos, D.; Vallianatos, F. Landslide susceptibility mapping by comparing the WLC and

WofE multi-criteria methods in the West Crete Island, Greece. *Environ. Earth Sci.* **2014**, *72*, 5197–5219.

46. Papathanassiou, G.; Valkaniotis, S.; Ganas, A.; Pavlides, S. GIS-based statistical analysis of the spatial distribution of earthquake-induced landslides in the island of Lefkada, Ionian Islands, Greece. *Landslides* **2013**, *10*, 771–783.

47. Chalkias, C.; Kalogirou, S.; Ferentinou, M. Landslide susceptibility, Peloponnese Peninsula in South Greece. *J. Maps***2014**, *10*, 211–222.

48. Tsangaratos, P.; Benardos, A. Estimating landslide susceptibility through an artificial neural network classifier. *Nat. Hazards* **2014**, *74*, 1489–1516.

49. Gemitzi, A.; Falakakis, G.; Eskioglou, P.; Petalas, C. Evaluating landslide susceptibility using environmental factors, fuzzy membership functions and GIS. *Glob. NEST J.* **2011**, *13*, 28–40.

50. Ferentinou, M.; Chalkias, C. Mapping mass movement susceptibility across Greece with GIS, ANN and statistical methods. *Landslide Sci. Pract.* **2013**, 321–327.

51. Sabatakakis, N.; Koukis, G.; Vassiliades, E.; Lainas, S. Landslide susceptibility zonation in Greece. *Nat. Hazards* **2013**,*65*, 523–543.

52. Gournellos, T.; Chalkias, C.; Tsagas, D. Landslide susceptibility zonation of Greece using fuzzy logic and GIS.*Geographies* **2006**, *12*, 114–126.

53. Hellenic Statistical Authority (ELSTAT). *Population and Housing Census: Resident Population*; ELSTAT: Piraeus, Greece, 2011.

54. Pnevmatikos, J.D.; Katsoulis, B.D. The changing rainfall regime in Greece and its impact on climatological means.*Meteorol. Appl.* **2006**, *13*, 331–345.

55. Koukis, G.; Tsiambaos, G.; Sabatakakis, N. Landslides movements in Greece: Engineering geologicalcharacteristics and environmental consequences. In Proceedings of the 7th International Symposium of Engineering Geology and Environment, Athens, Greece, 23–27 June 1997; Volume 1, pp. 789–792.

56. Hellenic Military Geographical Service (HMGS). *Aerial Photography of Various Areas of Greece, Scale 1:40,000*; HMGS: Athens, Greece, 1969.

57. Che, V.B.; Kervyn, M.; Suh, C.E.; Fontijn, K.; Ernst, G.G.J.; del Marmol, M.A.; Trefois, P.; Jacobs, P. Landslide susceptibility assessment in Limbe (SW Cameroon): A field calibrated seed cell and information value method. *Catena***2012**, *92*, 83–98.

58. De Guidi, G.; Scudero, S. Landslide susceptibility assessment in the Peloritani Mts. (Sicily, Italy) and clues for tectonic control of relief processes. *Nat. Hazards Earth Syst. Sci.* **2013**, *13*, 949–963.

59. Youssef, A.M.; Pradhan, B.; Pourghasemi, H.R.; Abdullahi, S. Landslide susceptibility assessment at Wadi Jawrah Basin, Jizan region, Saudi Arabia using two bivariate models in GIS. *Geosci. J.* **2015**, *19*, 449–469.

60. Cruden, D.M.; Varnes, D.J. Landslide types and processes. In *Landslides: Investigation and Mitigation, Special Report*; Turner, A.K., Schuster, R.L., Eds.; National Academy Press: Washington, DC, USA, 1996.

61. Fell, R. Landslide risk assessment and acceptable risk. *Can. Geotech. J.* **1994**, *31*, 261–272.

62. Magliulo, P.; Lisio, A.; Russo, F.; Zelano, A. Geomorphology and landslide susceptibility assessment using GIS and bivariate statistics: A case study in southern Italy. *Nat. Hazards* **2008**, *47*, 411–435.

63. Das, O.; Sonmez, H.; Gokceoglu, C.; Nefeslioglu, H. A study on the effect of the ground acceleration on landslide susceptibility maps. *Geophys. Res. Abstr.* **2010**, *12*, EGU2010-5885-1.

64. Yalcin, A. GIS-based landslide susceptibility mapping using analytical process and bivariate statistics in Ardesen (Turkey): Comparisons of results and confirmations. *Catena* **2008**, *72*, 1–12.

65. Hasekioğullari, G.D.; Ercanoglu, M. A new approach to use AHP in landslide susceptibility mapping: A case study at Yenice (Karabuk, NW Turkey). *Nat. Hazards* **2012**, *63*, 1157–1179.

66. Heymann, Y.; Steenmans, C.; Croisille, G.; Bossard, M. *CORINE Land Cover: Technical Guide*; Office for Official Publications of the European Communities: Luxembourg City, Luxembourg, 1994.

67. Carrara, A.; Cardinali, M.; Detti, R.; Guzzetti, F.; Pasqui, V.; Reichenbach, P. GIS techniques and statistical models in evaluating landslide hazard. *Earth Surf. Process. Landf.* **1991**, *16*, 427–445.

68. Institute of Geology and Mineral Exploration (IGME). *Geological Map of Greece, Scale 1:500,000*, 2nd ed.; IGME: Athens, Greece, 1983.

69. Underhill, J.R. Late Cenozoic deformation of the Hellenide foreland, western Greece. *Geol. Soc. Am. Bull.* **1989**, *101*, 613–634.

70. Ayalew, L.; Yamagishi, H. The application of GIS-based logistic regression for landslide susceptibility mapping in the Kakuda-Yahiko Mountains Central Japan. *Geomorphology* **2005**, *65*, 15–31.

71. Rahmani, M.; Hosieni, S.A.; Molla Aghajanzadeh, A.; Mashayekhan, A.; Mahalleh, S.M.M. Landslide hazard zonation using the MCE method (Case study: Agh Mashhad, Mazandaran Province, Iran). *Int. J. AgriSci.* **2013**, *3*, 240–244.

72. Ahmed, M.F.; Rogers, J.D.; Ismail, E.H. A regional level preliminary landslide susceptibility study of the upper Indus river basin. *Eur. J. Remote Sens.* **2014**, *47*, 343–373.

73. Sato, H.P.; Hasegawa, H.; Fujiwara, S.; Tobita, M.; Koarai, M.; Une, H.; Iwahashi, J. Interpretation of landslide distribution triggered by the 2005 Northern Pakistan earthquake using SPOT 5 imagery. *Landslides* **2007**, *4*, 113–122.

74. Owen, L.A.; Kamp, U.; Khattak, G.A.; Harp, E.L.; Keefer, D.K.; Bauer, M.A. Landslides triggered by the 8 October 2005 Kashmir earthquake. *Geomorphology* **2008**, *94*, 1–9.

75. Huade, G.; Wilson, J.L.; Makhnin, O. Geostatistical mapping of mountain precipitation incorporating autosearched effects of terrain and climatic characteristics. *J. Hydrometeorol.* **2005**, *6*, 1018–1031.

76. Arjun, C.R.; Kumar, A. Artificial neural network-based estimation of peak ground acceleration. *ISET J. Earthq. Technol.***2009**, *46*, 19–28.

77. Stournaras, G.; Tsimpidis, G.; Tsoumanis, P.; Yannatos, G.; Guillande, R. Instability Phenomena in the Neogenne Deposits of Northern Peloponnese. *Bull. Eng. Geol. Environ.* **1998**, *57*, 65–68.

78. Papathanassiou, G. Estimating slope failure potential in an earthquake prone area: A case study at Skolis Mountain, NW Peloponnesus, Greece. *Bull. Eng. Geol. Environ.* **2012**, *71*, 187–194.

79. Dounias, G.; Fikiris, I.F.; Kollios, A.; Kabbadas, M.; Marinos, P. The Nemea Landslide. Trigger Mechanism–Remedial Measures. In Proceedings of the 5th Panellenic Conference of Geotechnics and Geoenvironmental Engineering, Xanthi, Greece, 31 May–2 June 2006.

80. Papadopoulos, G.A.; Plessa, A. Magnitude-distance relations for earthquake-induced landslides in Greece. *Eng. Geol.* **2000**, *58*, 377–386.

81. Chousianitis, K.; Del Gaudio, V.; Sabatakakis, N.; Kavoura, K.; Drakatos, G.; Bathrellos, G.D.; Skilodimou, H.D. Assessment of earthquake-induced landslide hazard in Greece: From Arias Intensity to spatial distribution of slope resistance demand. *Bull. Seismol. Soc. Am.* **2016**.

82. Van Westen, C.J. Statistical landslide hazard analysis. In *ILWIS 2.1 for Windows Application Guide*; ITC Publication: Utrecht, The Netherlands, 1997; pp. 73–84.

83. Jenks, G.F. The data model concept in statistical mapping. *Int. Yearb. Cartogr.* **1967**, *7*, 186–190.

84. Lee, E.M.; Jones, D.K.C. *Landslide Risk Assessment*; Thomas Telford: London, UK, 2004; p. 161.

85. Malczewski, J. *GIS and Multicriteria Decision Analysis*; John Wiley & Sons: New York, NY, USA, 1999.

86. Remondo, J.; González-Díez, A.; Díaz de Terán, J.R.; Cendrero, A. Landslide susceptibility models utilizing spatial data analysis techniques. A case study from the Lower Deba Valley, Guipúzcoa (Spain). *Nat. Hazards* **2003**, *30*, 267–279.

87. Nandi, A.; Shakoor, A. Preparation of a landslide susceptibility map of summit county, Ohio, USA, using numerical models. In Proceedings of the 10th IAEG Congress, Nottingham, UK, 6–10 September 2006.

88. Frattini, P.; Crosta, G.; Carrara, A. Techniques for evaluating the performance of landslide susceptibility models. *Eng. Geol.* **2010**, *111*, 62–72.

89. Fawcett, T. An introduction to ROC analysis. *Pattern Recognit. Lett.* **2006**, *27*, 861–874.

90. Dragićević, S.; Lai, T.; Balram, S. GIS-based multicriteria evaluation with multiscale analysis to characterize urban landslide susceptibility in data-scarce environments. *Habitat Int.* **2015**, *45*, 114–125.

91. Faraji Sabokbar, H.; Shadman Roodposhti, M.; Tazik, E. Landslide susceptibility mapping using geographically-weighted principal component analysis. *Geomorphology* **2014**, *226*, 15–24.

92. Yang, Z.H.; Lan, H.X.; Gao, X.; Li, L.P.; Meng, Y.S.; Wu, Y.M. Urgent landslide susceptibility assessment in the 2013 Lushan earthquake-impacted area, Sichuan Province, China. *Nat. Hazards* **2015**, *75*, 2467–2487.

93. Van Westen, C.J.; Rengers, N.; Terlien, M.T.J.; Soeters, R. Prediction of the occurrence of slope instability phenomena through GIS-based hazard zonation. *Geol. Rundsch.* **1997**, *86*, 404–414.

94. Ayalew, L.; Yamagishi, H.; Maruib, H.; Takami, K. Landslides in Sado Island of Japan: Part II. GIS-based susceptibility mapping with comparisons of results from two methods and verifications. *Eng. Geol.* **2005**, *81*, 432–445.

95. Tien Bui, D.; Lofman, O.; Revhaug, I.; Dick, O. Landslide susceptibility analysis in the Hoa Binh province of Vietnam using statistical index and logistic regression. *Nat. Hazards* **2011**, *59*, 1413–1444.

96. Van Westen, C.J.; Rengers, N.; Soeters, R. Use of geomorphological information in indirect landslide susceptibility assessment. *Nat. Hazards* **2003**, *30*, 399–419.

97. Saboya, F., Jr.; da Glória Alves, M.; Dias Pinto, W. Assessment of failure susceptibility of soil slopes using fuzzy logic.*Eng. Geol.* **2006**, *86*, 211–224.

98. Bathrellos, G.D.; Gaki-Papanastassiou, K.; Skilodimou, H.D.; Papanastassiou, D.; Chousianitis, K.G. Potential suitability for urban planning and industry development by using natural hazard maps and geological—Geomorphological parameters. *Environ. Earth Sci.* **2012**, *66*, 537–548.

99. Rozos, D.; Skilodimou, H.D.; Loupasakis, C.; Bathrellos, G.D. Application of the revised universal soil loss equation model on landslide prevention. An example from N. Euboea (Evia) Island, Greece. *Environ. Earth Sci.* **2013**, *70*, 3255–3266.

3

# SPATIAL PATTERN ANALYSIS OF HEAVY METALS IN BEIJING AGRICULTURAL SOILS BASED ON SPATIAL AUTOCORRELATION STATISTICS

**Xiao-Ni Huo [1,3], Wei-Wei Zhang [1], Dan-Feng Sun [2], Hong Li [1,*], Lian-Di Zhou [1] and Bao-Guo Li [2]**

[1]Beijing Academy of Agriculture and Forestry, Beijing 100097, China

[2]Department of Land Resources and Management, College of Natural Resources and Environment Science, China Agricultural University, Beijing 100193, China

[3]Department of Environmental Engineering, Taiyuan University, Taiyuan 030009, China

## ABSTRACT

This study explored the spatial pattern of heavy metals in Beijing agricultural soils using Moran's I statistic of spatial autocorrelation. The global Moran's I result showed that the spatial dependence of Cr, Ni, Zn, and Hg changed with different spatial weight matrixes, and they had significant and positive global spatial correlations based on distance weight. The spatial dependence of the four metals was scale-dependent on distance, but these scale effects existed within a threshold distance of 13 km, 32 km, 50 km, and 29 km, respectively for Cr, Ni, Zn, and Hg. The maximal spatial positive correlation range was 57 km, 70 km, 57 km, and 55 km for Cr, Ni, Zn, and Hg, respectively and these were not affected by sampling density. Local spatial autocorrelation analysis detected the locations of spatial clusters and spatial outliers and revealed that the pollution of these four metals occurred in significant High-high spatial clusters, Low-high, or even High-low spatial outliers. Thus, three major areas were identified and should be receiving more attention: the first was the northeast region of Beijing, where Cr, Zn, Ni, and Hg had significant increases. The second was the southeast region of Beijing where wastewater irrigation had strongly changed the content of metals, particularly of Cr and Zn, in soils. The third area was the urban fringe around city, where Hg showed a significant increase.

## KEYWORDS

heavy metals; spatial pattern; Moran's I statistic; Beijing agricultural soils

## 1. INTRODUCTION

Heavy metals often accumulate to excess in agricultural soils due to natural processes and anthropogenic activities. Since these can cause adverse effects on the environment, heavy metal soil pollution is an urgent problem. Successful assessment and remediation of heavy metal pollution in soils will depend on the understanding of their spatial variability and the relationships between heavy metals and the factors leading to pollution.

Both geostatistics and spatial autocorrelation statistics (Moran's I) are the methods used for exploring the spatial pattern of heavy metals. Generally, geostatistics is widely used [1–4], but it cannot determine whether the spatial correlation is positive or negative, and the significance of spatial dependence. In addition, geostatistics cannot detect spatial outliers, whose values obviously different from the values of their surrounding location, can make the semivariogram erratic.

Compared with geostatistics, global Moran's I can be used to estimate the strength and determine the positive or negative of the spatial correlation of a variable, local Moran's I can be used to identify spatial clusters and spatial outliers of a variable [5–7]. Furthermore, the significance of the spatial correlation can be tested [8,9]. Although spatial autocorrelation was defined decades ago and has been widely used in many studies [8,10,11], only the global Moran's I statistic has been used for the spatial variability of heavy metals [10]. Reports on analyzing multi-scale spatial variability of heavy metals in soils using the Moran's I statistic have not yet been published, particularly in relation to the local Moran's I statistic for understanding the structure of spatial dependence.

For heavy metals in Beijing soils, there are some related studies. However, the emphasis of such studies just covered the urban-rural transition zone, periurban, and rural zones, and/or had a relatively few sampling points [12–14]. Thus, in order to reflect the whole and representative real situation of heavy metals in whole Beijing agricultural soils, Huo *et al.* further assessed the spatial variability of heavy metals with a total of 1,018 samples covering the entire Beijing agricultural soils using geostatistics [15]. Considering the advantages of the Moran's I statistic, it can help to identify the more characteristics of the spatial pattern of heavy metals, because understanding the spatial pattern is the basis for environmental quality assessment and soil remediation.

Therefore, the main objectives of this research were: (1) to explore the spatial variability of heavy metals in Beijing agricultural soils using the spatial autocorrelation statistic, including the influence of different weight matrixes and sampling density on spatial autocorrelation; and (2) to reveal the local spatial patterns of heavy metals in Beijing agricultural soils in order to identify the potential risk areas for heavy metal pollution and its possible causes.

# 2. MATERIALS AND METHODS

## 2.1. Study Area

Beijing municipality is located in the northwest region of the north China plain, between longitude 115°25′–117°30′E and latitude 39°28′–41°05′N, and covers an estimated area of $1.6 \times 10^4$ km$^2$. The elevation of Beijing ranges from 2,250 m in the northwest to 9 m in the southeast. Mountains cover approximately 62% of the entire area and plains cover the remaining 38% (Figure 1). The area has a temperate continental monsoonal climate with an average annual temperature of 11.8 °C (average maximum 26 °C in July and average minimum −5 °C in January). The mean annual precipitation in the area is 470–660 mm, with approximately 60% of the precipitation occurring in July and August. The average annual evaporation is 1,800–2,000 mm. The area is the source of five large rivers, the Yongding, Chaobai, Beiyun, Jiyun and Daqing (Figure 1). Average annual runoff is about $1.8 \times 10^9$ m$^3$ but has decreased to $1.3 \times 10^9$ m$^3$ since the end of the last century. The Yongding River provides water mainly for industry, the Chaobai River mainly for resident living, and the Beiyun River plays a dual role in wastewater drainage from industry and human living activities.

The primary types of agricultural soil in the area include Cab Ustic Luvisols, Hap Ustic Luvisols and Och Aquic Cambisols. According to the second agricultural census in 2006, cultivated land occupied $2.32 \times 10^5$ hm$^2$, and orchard land $1.22 \times 10^5$ hm$^2$, together covering 21% the total area of Beijing. The fertilizer input was 517.5 kg/hm$^2$, about 1.44 times of the national average level, and the pesticide inputs were 8.11 kg/hm$^2$, or about 2.33 times of the country level. The overuse of fertilizer and pesticides has imposed heavy environmental pressures on the area. The metal mineral resources were primarily Fe, Ag, Zn, and Pb elements, and ironstone was exploited on a large-scale, mainly distributed in the Huairou district and Miyun county (Figure 1). In addition, there had been white marble exploration in southwest Beijing (Figure 1). The mineral resource exploration and industry development in Beijing also had negative environmental effects on the area.

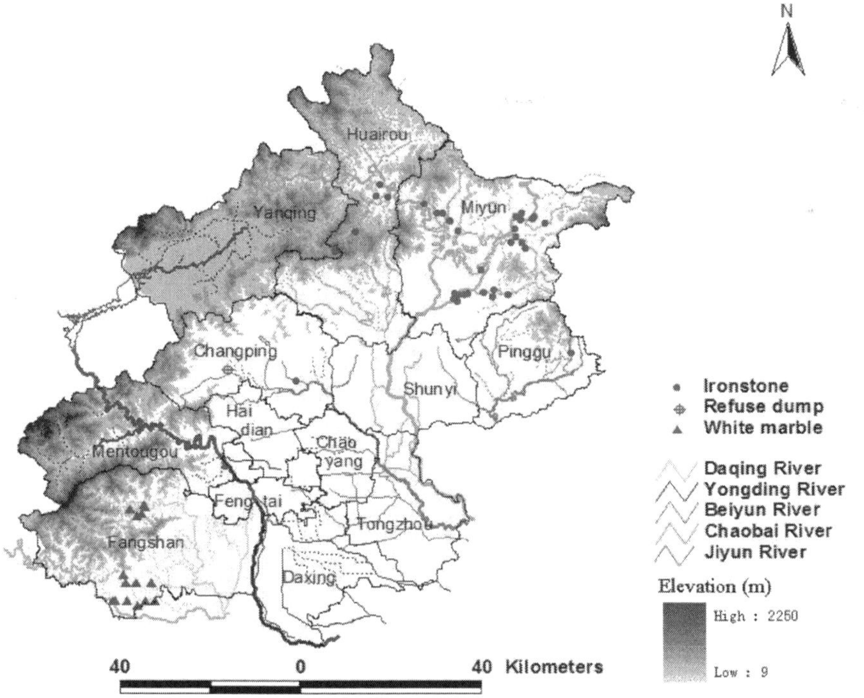


**Figure 1.** The study area.

## 2.2. Soil Sampling and Analysis

To investigate the pollution status of heavy metals in Beijing agricultural areas, a large-scale soil sampling project was conducted after the crop harvest in the Autumn of 2006. According to the agricultural land distribution and land use type maps of Beijing, a non-uniform distribution of the stratified sampling technique was adopted to collect samples and ensure the representativeness of sample. The sampling process was divided into three steps to collect a total of 1,018 samples. First, 231 soil samples were collected from the agricultural soils in the entire study area, with uniform sampling being the low sampling density (C). Secondly, another 360 soil samples were added from areas with more agricultural soils to create the medium sampling density (M). Third, 427 soil samples were further collected on the basis of the two previous samplings and the agricultural soils to make a high sampling density (F). General ideal is that more areas of agricultural land more sampling points. Figure 2 shows the distribution of soil samples at the three sampling densities.

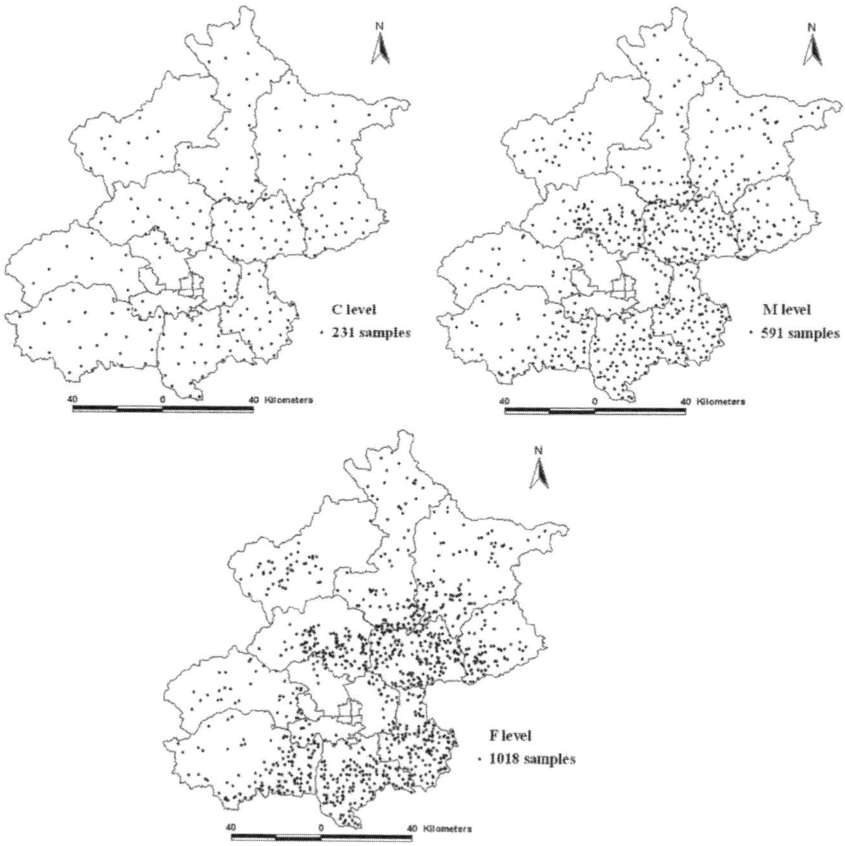


**Figure 2.** Distribution of soil samples at the three densities.

For each sample, five surface soil (0~20 cm) sites were sampled within 10 × 10 m square areas and then mixed. Global Positioning System was used to precisely locate each sampling position (latitude and longitude);

and a total of 1 kg of mixed soil per sample was collected. All soil samples were collected using a stainless steel spade and a scoop made from bamboo and then stored in polyethylene bags.

The soil samples were air-dried, crushed in an agate mortar, and then passed through a 100-mesh nylon sieve. The concentrations of eight heavy metals, including Cr, Ni, Cu, Zn, As, Cd, Pb, and Hg, were analyzed in the soil samples following the Chinese Environmental Quality Standard for Soils (GB15618-1995).

## 2.3. Global Spatial Autocorrelation

Spatial autocorrelation is an assessment of the correlation of a variable with reference to its spatial location and it deals with the attributes and the locations of the spatial features [16]. Moran's I is a popular test statistic for spatial autocorrelation.

Global Moran's I is a global test statistic for spatial autocorrelation, which is based on cross-products for measuring attribute association. It is calculated for *n* observations on a variable *x* at locations of *i* and *j*, as follows [17]:

$$I = \frac{n}{\sum_{i=1}^{n}\sum_{j=1}^{n} w_{ij}} \cdot \frac{\sum_{i=1}^{n}\sum_{j=1}^{n} w_{ij}\left(x_i - \bar{x}\right)\left(x_j - \bar{x}\right)}{\sum_{i=1}^{n}\left(x_i - \bar{x}\right)^2} \qquad i \neq j \tag{1}$$

where *n* is the number of observations of the whole region, $x_i$ and $x_j$ are the observations at locations of *i* and *j*, $\bar{x}$ is the mean of *x*, and $w_{ij}$, an element of spatial weights matrix *w*, is the spatial weight between locations of *i* and *j*.

The weight matrix depicts the relationship between an element and its surrounding elements. The weights are based on contiguity relations or distance. In a weight matrix based on contiguity, a value unequal to zero in the matrix represents pairs of elements with a certain contiguity relation and a zero represents pairs without contiguity relation. The rook case and queen case are the typical examples for contiguity relation. The first takes only four neighbours into account with common boundaries, and the latter takes into account all eight surrounding cells, including common boundaries and common corners. In a distance-based weight matrix, a threshold distance is specified such that all locations within the given distance are considered to be "neighbors". Alternatively, the k-nearest neighbor weight matrix is also based on distance, which is computed as the distance between a point and the number (k) of nearest neighbor points.

The value of Moran's I will vary between 1 and −1. A higher positive Moran's I implies that values in neighboring positions tend to cluster, while a low negative Moran's I indicates that high and low values are interspersed. When Moran's I is near zero, there is no spatial autocorrelation, meaning that the data are randomly distributed [8,9].

The Moran scatter plot describes the distribution of all observation values (x-axis) in relation to their neighbors (y-axis). Observations in the lower left (Low–low) and upper right (High–high) quadrant represent potential spatial clusters (values surrounded by similar neighbors), whereas observations in the upper left (Low–high) and lower right (High–low) suggest potential spatial outliers (values surrounded by dissimilar neighbors) [18]. The slope of the scatter plot corresponds to the value of global Moran's I, therefore, the scatter plot can help to visualize the strength of the overall spatial autocorrelation, as well as the complexity of spatial autocorrelation types in the study area. However, the Moran scatter plot does not indicate the significance of spatial autocorrelation types and the locations of the clusters or outliers.

## 2.4. Local Spatial Autocorrelation

Local Moran's I is a local test statistic for spatial autocorrelation, and identifies the autocorrelation between a single location and its neighbors. It is computed as follows:

$$I_i = \frac{n\left(x_i - \bar{x}\right)\sum_{j=1}^{n} w_{ij}\left(x_j - \bar{x}\right)}{\sum_{i=1}^{n}\left(x_i - \bar{x}\right)^2} \qquad (2)$$

The notations in Equation (2) are as described for Equation (1), but the corresponding values are from the local neighboring region.

A map showing locations with significant Local Moran statistics, classified by types of spatial correlation, is defined as a Local Indicators of Spatial Autocorrelation (LISA) cluster map [5]. Four categories of local spatial autocorrelation are distinguished; two of these suggest clusters and two of these suggest outliers. The LISA map provides information about which clusters/outliers are statistically significant and indicates the variability and distribution of the spatial correlation types in the study area. Therefore, LISA is an important tool for detecting "interesting" locations and assessing the extent to which the spatial distribution exhibits "spatial heterogeneity" [5].

## 2.5. Data Treatment with Computer Software

Soil samples were stored using the ArcView 3.2 software to create a spatial database. The Thiessen polygons of samples were created and all spatial analysis including global and local spatial autocorrelation was carried out using Geoda095i software [18].

# 3. RESULTS AND DISCUSSION

## 3.1. Influence of Different Weight Matrixes on Spatial Autocorrelation

Eight different spatial weight matrixes were used to impose a neighborhood structure on the 1,018 samples and assess the spatial autocorrelation of eight heavy metals. The results of global Moran's I values are given in Table 1. The significance of Moran's I were tested ($p$

< 0.05). Cr, Ni, Zn, As, Cd, and Hg showed significant and positive spatial correlations on all spatial weights, but As and Cd had low Moran's I values close to 0. There were no significant spatial correlations for Cu on any spatial weights. Except for the 4-nearest neighbors weight, Pb had spatial significant correlations on other types of weights, but the values of Moran's I were near 0.

Generally, the numbers of neighbours using the queen criterion will be equal to or greater than that using the rook criterion. However, Table 1 shows that the global Moran's I values based on the first order rook and queen weights were equivalent. This may be because they had the same connectivity histogram. This indicated that the direct neighboring relations were not affected by the direction of four neighbors or eight neighbors. Spatial autocorrelation coefficients generally decreased with the increase of the number (k) of nearest neighbor points, following the rule of the farther the distance, the less attribute similarity (Table 1). The metals had relatively high Moran's I values with the 4 km distance band spatial weight matrix. Furthermore, for irregular samples, the reasonable weight matrix is the distance-based weight matrix. Therefore, the subsequent spatial correlation analysis was calculated using this weight matrix based on distance, and focused on the four elements, Cr, Ni, Zn, and Hg, which had significant and positive spatial correlations.

Because the selection of spatial weigh was empirical, as well as the same weight matrix under a certain distance limit was assigned to all points, there may be had a certain impact on the spatial autocorrelation of the heavy metals. If the spatial weights based on decay distance were designed, the results of the influence of spatial weight on spatial autocorrelation of heavy metals may be more reasonable.

Figure 3 shows the Moran scatter plots of Cr, Ni, Zn, and Hg with 1,018 samples, in which the horizontal axis was the standardized value of heavy metal concentration and the vertical axis was the standardized value of the neighboring heavy metal concentration. A large part of the samples of the four metal elements mainly clustered in the left lower and right upper quadrants, indicating that a positive spatial autocorrelation dominated the overall spatial pattern.

**Table 1.** Global spatial autocorrelation coefficient (global Moran's I value) based on different spatial weighs for heavy metals.

| Spatial weights | Cr | Ni | Cu | Zn | As | Cd | Pb | Hg |
|---|---|---|---|---|---|---|---|---|
| First order rook contiguity | 0.495 * | 0.317 * | 0.026 | 0.282 * | 0.077 * | 0.119 * | 0.051 * | 0.290 * |
| First order queen contiguity | 0.495 * | 0.317 * | 0.026 | 0.282 * | 0.077 * | 0.119 * | 0.051 * | 0.290 * |
| 4-nearest neighbors | 0.541 * | 0.327 * | 0.006 | 0.287 * | 0.071 * | 0.124 * | 0.036 | 0.283 * |
| 5-nearest neighbors | 0.521 * | 0.320 * | 0.018 | 0.277 * | 0.080 * | 0.125 * | 0.047 * | 0.275 * |
| 6-nearest neighbors | 0.513 * | 0.321 * | 0.018 | 0.265 * | 0.084 * | 0.122 * | 0.049 * | 0.271 * |
| 7-nearest neighbors | 0.498 * | 0.315 * | 0.018 | 0.253 * | 0.078 * | 0.118 * | 0.045 * | 0.263 * |
| 8-nearest neighbors | 0.486 * | 0.309 * | 0.018 | 0.246 * | 0.073 * | 0.119 * | 0.047 * | 0.258 * |
| 4km distance band | 0.472 * | 0.318 * | 0.024 | 0.293 * | 0.090 * | 0.127 * | 0.056 * | 0.272 * |

* Significant at the 0.05 level.

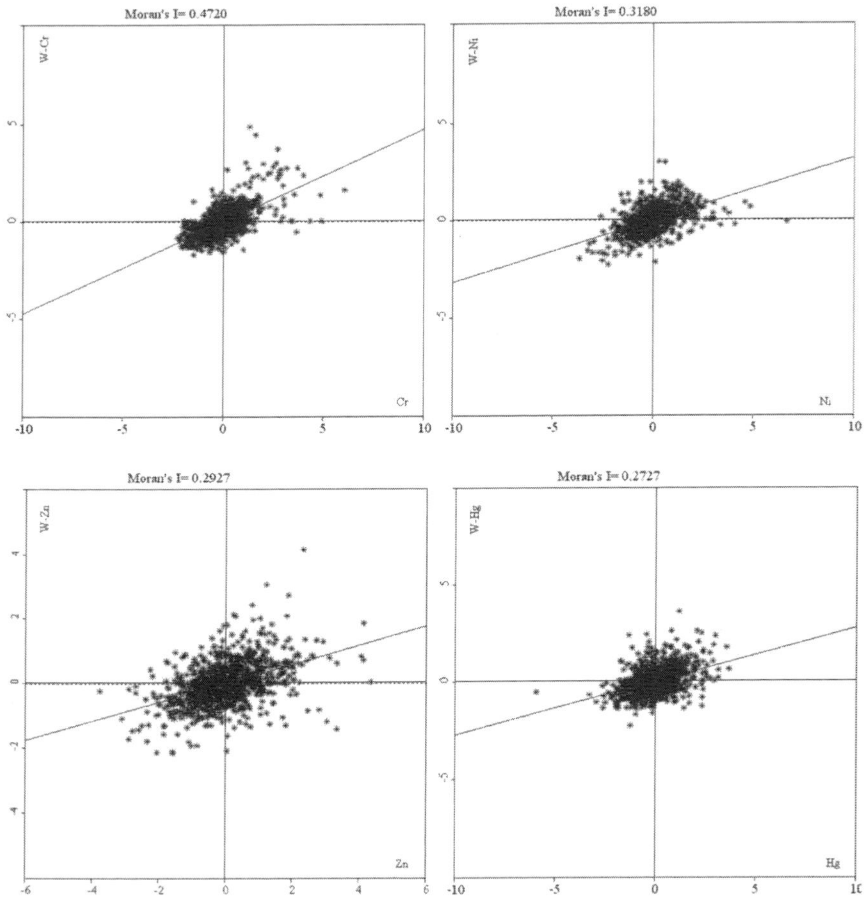


**Figure 3.** Moran scatter plot for Cr, Ni, Zn, and Hg.

There was also a certain part of the samples in the right lower and left upper quadrants, indicating that negative spatial autocorrelation could not be neglected. With the decrease in spatial autocorrelation coefficients, the scatter plot tended to became further disaggregated, and these samples were far from the Moran's I regression line and strongly influenced the global spatial autocorrelation, particularly for Cr and Zn, indicating some local nonstationarity (Figure 3). Consequently, the variability of their spatial patterns should be considered.

## 3.2. The Effect of Sampling Density on Spatial Autocorrelation

Moran's I values can be plotted against distance classes, called a spatial correlogram [8]. Figure 4 gives the spatial correlograms for Cr, Ni, Zn, and Hg produced with a weight matrix based on distance at three sampling densities. The Moran's I of Cr, Ni, Zn, and Hg all initially increased as a peak with the increase of distance, then dropped down to 0 with further increase of the distance at three density levels. This revealed, at all three sampling densities, the four heavy metals displayed stronger spatial dependence initially as the distance expanded to include more close points, whereas their spatial dependence decreased as the distance further increased to include more distant points. Normally, the distance where the 0 value first appears is considered as the maximal spatial positive correlation range, which was about 57 km, 70 km, 57 km, and 55 km for Cr, Ni, Zn, and Hg, respectively (Figure 4).

Moran's I for Cr had a distinct difference within 13 km, then the difference disappeared for the three density levels [Figure 4(a)]. Similarly, Ni, Zn, and Hg also had significant differences when the distance was less than 32 km, 50 km, and 29 km, respectively [Figure 4(b), (c), (d)]. This indicated that sampling density had an effect on the spatial dependence, but the effect was no longer obvious beyond a certain distance. Consequently, with the increase of the distance, the far neighbors gradually imposed on the spatial dependence instead of the sampling density, and sampling density had no influence on the maximal spatial positive correlation ranges of the four metals.

Figure 4(a) shows that the peak of the spatial correlogram for Cr was highest at the F level at about 6 km characteristic distance. As the sampling density decreased, the peak decreased, but the distance increased as the peak became greater. For Zn and Ni, the peaks at the F and M level were close about 4 km, and somewhat higher at the M level, while that at the C level was significantly low and fluctuant [Figure 4(b), (c)]. For Hg, the highest peak was at the C level about 6 km, and the peaks at the F, M level were about 4 km, but the peak values were relatively low [Figure 4(d)]. Therefore, Cr should adopt a high-density level, Hg can adopt a low-density level, and Zn and Ni can use a medium-density level for global spatial autocorrelation analysis, in order to reduce the sample numbers. In addition, the distances in which the peaks appeared at three sampling densities should be adopted for the spatial dependence analysis.

The spatial correlogram analysis revealed that Cr, Ni, and Zn had the similar sampling density effect. For Cr, Ni, and Zn, the higher sampling density enhanced the spatial dependence. In contrast, it may be the existence of extreme value, the spatial dependence of Hg at higher sampling density became weaker. These scale effects existed within a certain distance. Moreover, the spatial correlogram can help to find the maximal spatial positive correlation range and the suitable neighborhoods for spatial dependence analysis. Therefore, LISA was further used to identify the detail spatial variability for the four metals at the F density level.

## 3.3. Local Indicators of Spatial Association (LISA)

The LISA indicates the spatial variability details. The distance where the spatial dependence of Cr, Ni, Zn, and Hg was the strongest at F density level is selected to reveal the local spatial pattern (High-high, Low-low, High-low, Low-high and no significance). The distance weight matrix was 6 km, 4 km, 4 km, and 4 km for Cr, Ni, Zn, and Hg, respectively.

Although the four heavy metals had a significant global spatial positive correlation, more than half of the samples of the four metals had no significant spatial pattern (Table 2). Table 2 shows 36.5% Cr, about 20% of the other metals samples belonged to the significant spatial clusters, which were the largest spatial pattern. As the trace elements in soils, Low-low pattern of the four heavy metals dominated the overall spatial pattern, as well as High-high patterns that were more than half that of the Low-low pattern. About 10% samples of the metals also were significant spatial outliers, and the Low-high pattern for Ni, Zn, and Hg was overwhelming (Table 2). These significant spatial patterns can indicate strong ongoing enrichment processes of the four heavy metals in Beijing agricultural soils.

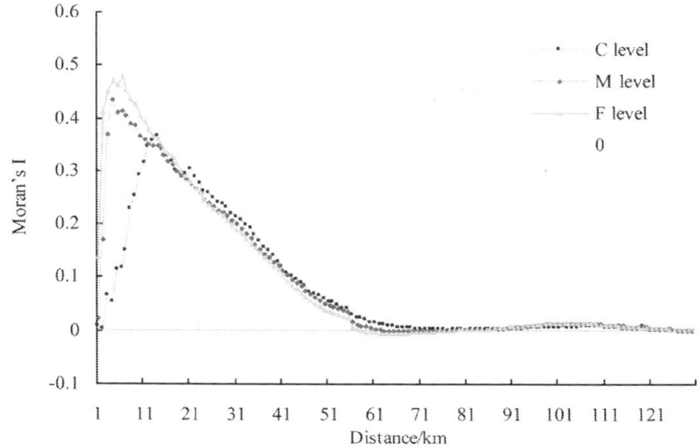


(**a**) The spatial correlograms of Cr at three sampling density.

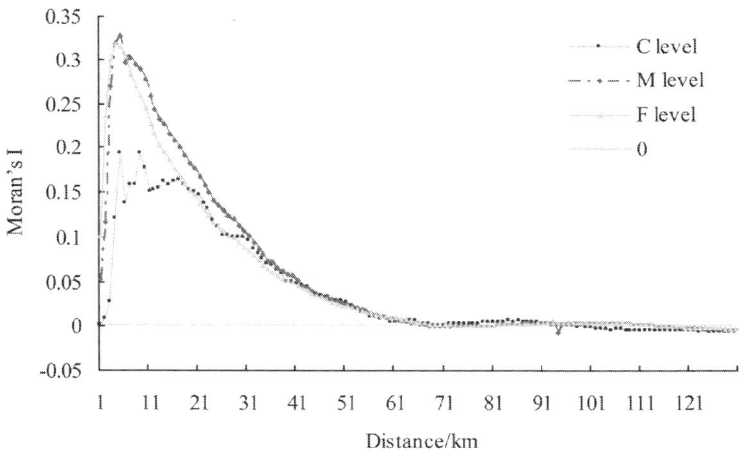


(**b**) The spatial correlograms of Ni at three sampling density.

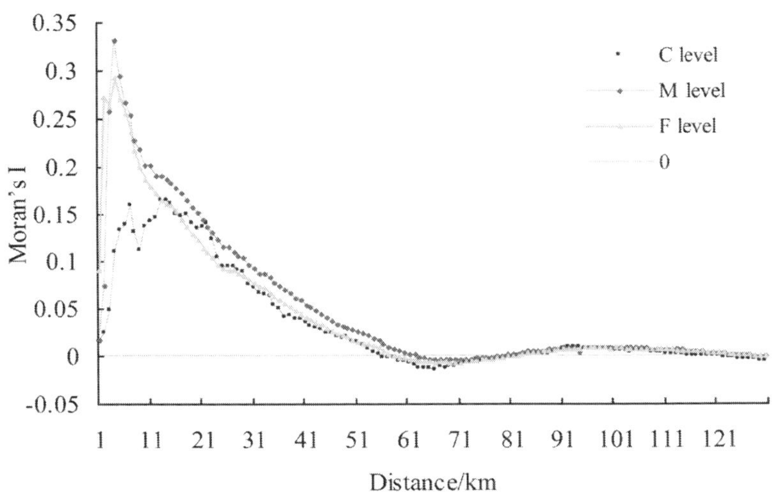

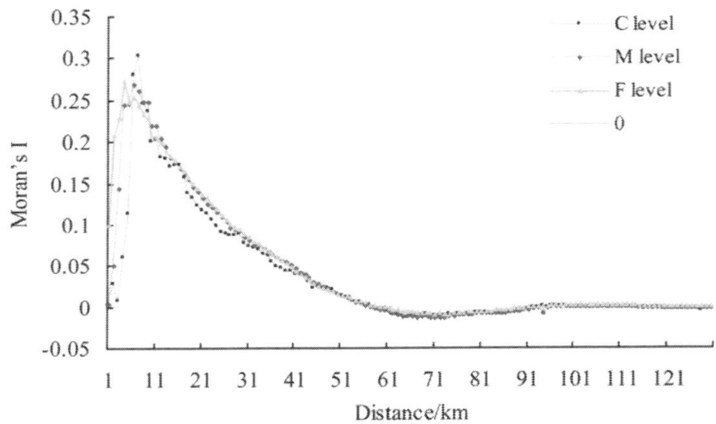


(**d**) The spatial correlograms of Hg at three sampling density.

**Figure 4.** The spatial correlograms of the metals at three sampling density.

**Table 2.** Sample percent of local spatial pattern types of LISA analysis (%).

| Types of spatial autocorrelation | Cr | Ni | Zn | Hg |
|---|---|---|---|---|
| No significance | 56.09 | 69.94 | 66.70 | 67.78 |
| High-high | 14.34 | 7.07 | 8.74 | 9.63 |
| Low-low | 22.20 | 12.48 | 13.46 | 11.30 |
| Low-high | 3.05 | 7.96 | 7.56 | 8.35 |
| High-low | 4.32 | 2.55 | 3.54 | 2.95 |

The samples with Cr pollution were only 0.69% and all of these occurred in the significant High-high spatial pattern. Zn pollution was less (only 0.1% samples) in the no significant spatial pattern type (Table 3). Hg and Ni represented a relatively high percentage of polluted samples, and about 6.2% of the samples were polluted by Hg. Among these, 2.26% of the samples were in significant High-high spatial clusters, 0.49% in significant Low-high, and only 0.10% in significant High-low spatial outliers (Table 3). For Ni, about 3.93% of the samples were polluted and 0.79% of the samples were in significant High-high spatial clusters, and 0.49% in significant Low-high spatial outliers (Table 3).

**Table 3.** Sample pollution status distribution in local spatial pattern types (%).

| Heavy metals | Pollution status | Types of spatial autocorrelation | | | | |
|---|---|---|---|---|---|---|
| | | No significance | High-high | Low-low | Low-high | High-low |
| Cr | Polluted | | 0.69 | | | |
| | Unpolluted | 56.09 | 13.65 | 22.20 | 3.05 | 4.32 |
| Ni | Polluted | 2.65 | 0.79 | | 0.49 | |
| | Unpolluted | 67.29 | 6.29 | 12.48 | 7.47 | 2.55 |
| Zn | Polluted | 0.10 | | | | |
| | Unpolluted | 66.60 | 8.74 | 13.46 | 7.56 | 3.54 |
| Hg | Polluted | 3.44 | 2.26 | | 0.49 | 0.10 |
| | Unpolluted | 64.34 | 7.37 | 11.30 | 7.86 | 2.85 |

Compared with the spatial randomness, the significant spatial patterns of these heavy metals demonstrated that the underlying enrichment processes were more stable, and as such, would present more difficulties for their remediation. In addition, in soil heavy metal evaluation, the outliers may represent the potential pollution areas, such as Ni and Hg in this study. If further spatial interpolation will be produced, the outliers cannot be deleted arbitrarily and should adopt more complicated geostatistics approach.

The LISA map can further detect the locations of the interesting spatial patterns for heavy metals. The northeast region was strongly influenced by High-high pattern of Cr, Ni [Figure 5(a), (b)], where some iron mines are distributed (Figure 1). The second hot region was the southeast Beijing, in which Cr, Ni, and Zn displayed a significant High-high pattern [Figure 5(a), (c)]. Moreover, High-low outliers of the three metals were mainly distributed in this region, and were found near Low-low clusters, indicating that anthropogenic activities had begun to change the Low-low pattern. In areas downstream of Beijing city, there had been long-term wastewater irrigation history in southeast Beijing, which led to heavy metal contamination [19]. The third interesting area was the urban fringe, particularly the northern and eastern parts, where clusters of High-high Hg were covered [Figure 5(d)]. Emissions from human activities such as the combustion of fossil fuels, the burning of coal, industrial boilers, and petroleum refineries had resulted in significant increases in the emission of Hg in and around urban areas [20,21]. As well, high concentrations of Cr, Ni, and Zn were clustered in and around the refuse dump in the southern Changping district, where landfill and waste incineration were the potential pollution sources (Figure 1, Figure 5).

However, because the LISA map was generated based on the soil samples, the boundaries between different spatial pattern types are easily confused and unreadable. In future research, the method of zoning should be introduced for improving to distinguish the spatial pattern types of boundaries. Such as, LISA clusters map, with other possible driving factors maps in GIS software, can also quantify their spatial relationships to confirm and refine their effects. Their distributions could be used to delineate the potential monitoring and remediation zones. Moreover, these zones can assist in developing measures and policies that can be responsive to the spatial variations and pollution processes.

# 4. CONCLUSIONS

Compared with geostatistics, although spatial autocorrelation analysis cannot be used to estimate unobserved points, global and local Moran's I analysis have their advantages. The strength and significance of spatial dependence could be easily compared and tested. Moreover, the spatial correlogram can describe the changes of spatial dependence with distance, which indicate the maximal spatial positive correlation range and the suitable neighborhoods for spatial dependence analysis. In addition, local Moran's I analysis can help to detect spatial outliers or hot spots. Therefore, it is an effective exploratory spatial analysis technique for regional variables.

This study explored spatial pattern of heavy metals in the entire Beijing agricultural soils based on 1,018 samples collected in 2006 using Moran's I analysis. The four elements Cr, Ni, Zn, and Hg had significant and positive spatial correlations with large Moran's I values. The Moran's I values were affected by spatial weight matrix. Cr, Ni, and Zn had similar sampling density effects. Higher sampling density enhance the spatial dependence, whereas, higher sampling density may reduce the spatial dependence for Hg. In addition, the maximal spatial positive correlation ranges of the four metals did not change at different sampling density. The global Moran's I of the four metals was scale-dependent on distance, initially taking stronger spatial dependence with the increase of the distance, then becoming weaker with further expansion of the distance. It is worth noting that the scale effect existed in a certain distance.

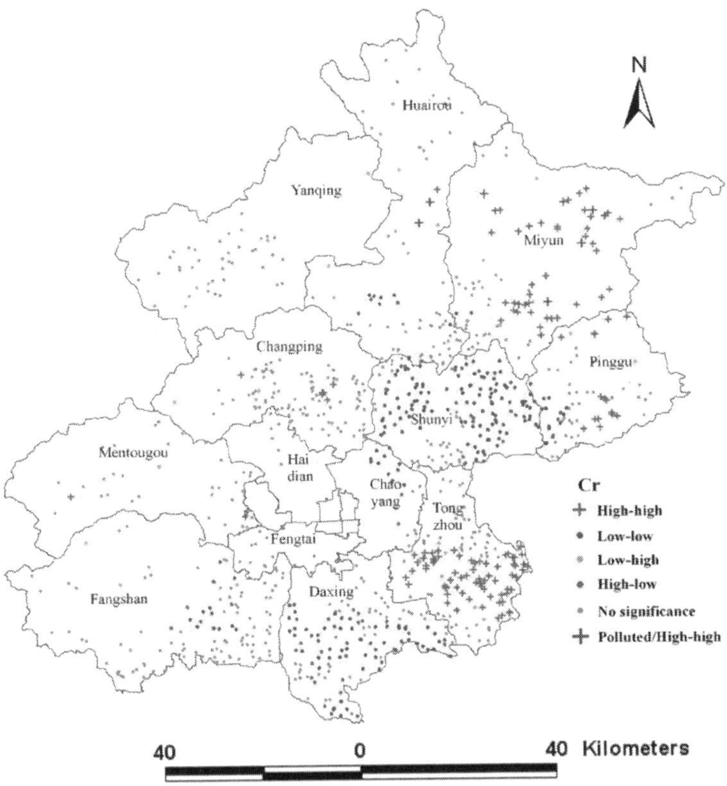


(a) LISA cluster map for Cr.

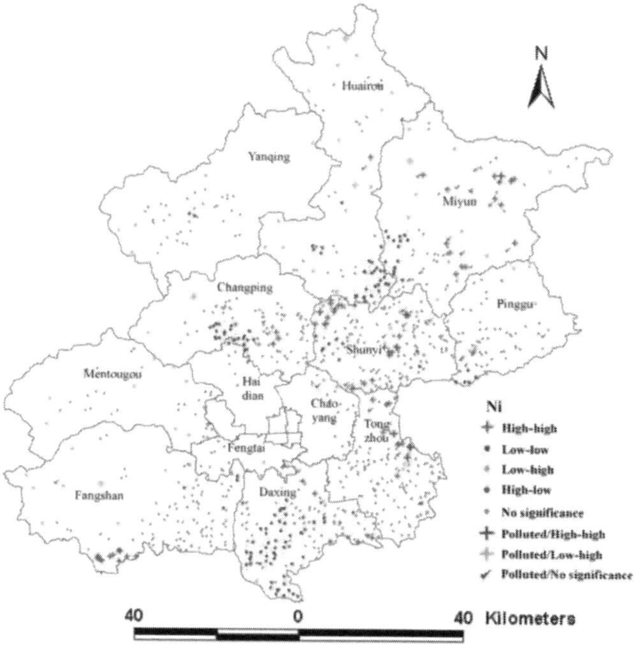


(**b**) LISA cluster map for Ni.

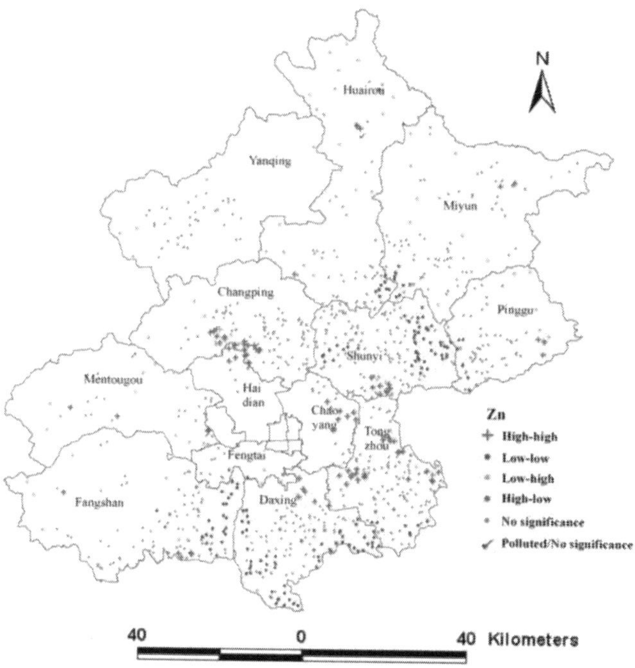


(**c**) LISA cluster map for Zn.

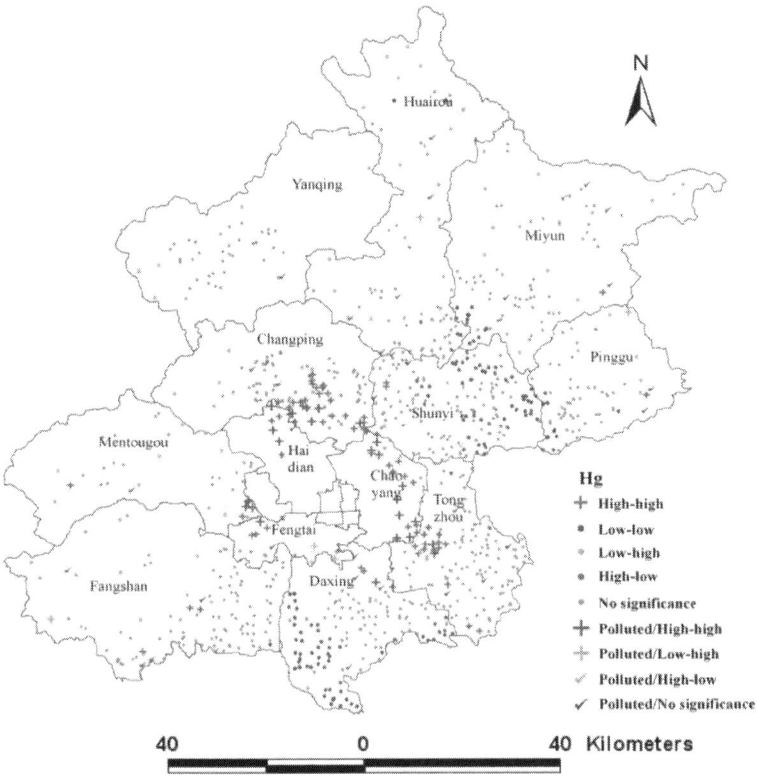


(d) LISA cluster map for Hg.

**Figure 5.** LISA cluster maps for heavy metals.

The local spatial autocorrelation analysis revealed that the four metals all had important High-high pattern, and Low-high and High-low spatial outliers, indicating strong enrichment processes for the four heavy metals in Beijing agricultural soils. Thus, these areas should be receiving more attention: the northeast and southeast region of Beijing, where significant increases in Cr, Zn, Ni, and Hg occurred, as well as the urban fringe around city where Hg showed a significant increase. The global Moran's I was proved to be a useful measure of overall clustering, while the local Moran's I was an important tool for detecting local spatial patterns for possible polluting areas or interesting patterns for further monitoring. Therefore, spatial autocorrelation analysis can be a useful method to explore the spatial pattern of heavy metals.

## ACKNOWLEDGMENTS

The research was supported by the National Key Technology R&D Program (Grant Nos. 2006BAD10A06 and 2004BA617B04) from the China Ministry of Sciences and Technology and Beijing agricultural scientific programs support. The authors are grateful to the Agricultural Environmental Monitoring Station of Beijing for their soil sampling and analysis.

## REFERENCES

1. McGrath, D; Zhang, CS; Carton, OW. Geostatistical analyses and hazard assessment on soil lead in Slivermines areas, Ireland. *Environ. Pollut* **2004**, *127*, 239–248.

2. Franco, C; Soares, A; Delgado, J. Geostatistical modelling of heavy metal contamination in the topsoil of Guadiamar river margins (S Spain) using a stochastic simulation technique. *Geoderma* **2006**, *136*, 852–864.

3. Liu, XM; Wu, JJ; Xu, JM. Characterizing the risk assessment of heavy metals and sampling uncertainty analysis in paddy field by geostatistics and GIS. *Environ. Pollut* **2006**, *141*, 257–264.

4. Hu, KL; Li, H; Li, BG; Huang, YF. Spatial and temporal patterns of soil organic matter in the urban-rural transition zone of Beijing. *Geoderma* **2007**, *141*, 302–310.

5. Anselin, L. Local indicators of association-LISA. *Geogr. Anal* **1995**, *27*, 93–115.

6. Boots, B. Local measures of spatial association. *Ecoscience* **2002**, *9*, 168–176.

7. Dale, MR; Fortin, MJ. Spatial autocorrelation and statistical tests in ecology. *Ecoscience* **2002**, *9*, 162–167.

8. Overmars, KP; de Koning, GHJ; Veldkamp, A. Spatial autocorrelation in multi-scale land use models. *Ecol. Model* **2003**,*164*, 257–270.

9. Zhang, CS; McGrath, D. Geostatistical and GIS analyses on soil organic carbon concentrations in grassland of southeastern Ireland from two different periods. *Geoderma* **2004**, *119*, 261–275.

10. Zhang, CS; Zhang, S; He, JB. Spatial distribution characteristics of heavy metals in the sediments of Changjiang River system-Spatial autocorrelation and fractal methods. *Acta Geogr Sinica* **1998**, *53*, 87–96. (in Chinese).

11. Anselin, L. Spatial effects in econometric practice in environmental and resource economics. *Am. J. Agric. Econ* **2001**,*83*, 705–710.

12. Hu, KL; Zhang, FR; Li, H; Huang, F; Li, BG. Spatial patterns of soil heavy metals in urban-rural transition zone of Beijing. *Pedosphere* **2006**, *16*, 690–698.

13. Tan, MZ; Xu, FM; Chen, J; Zhang, XL; Chen, JZ. Spatial prediction of heavy metal pollution for soils in pre-urban Beijing, China based on fuzzy set theory. *Pedosphere* **2006**, *16*, 545–554.

14. Wu, S; Xia, XH; Lin, CY; Chen, X; Zhou, CH. Levels of arsenic and heavy metals in the rural soils of Beijing and their changes over the last two decades (1985–2008). *J. Hazar. Mater* **2010**, *179*, 860–868.

15. Huo, XN; Li, H; Sun, DF; Zhou, LD; Li, BG. Multi-scale spatial structure of heavy metals in agricultural soils in Beijing. *Environ. Monit. Assess* **2010**, *164*, 605–616.

16. Goodchild, MF. *Spatial Autocorrelation*; CATMOG 47; Geobooks: Norwich, UK, 1986; pp. 6–25.

17. Cliff, AD; Ord, JK. *Spatial Processes: Models and Applications*; Pion: London, UK, 1981.

18. Anselin, L; Syabri, I; Kho, Y. GeoDa: An Introduction to spatial data Analysis. *Geogr. Anal* **2006**, *38*, 5–22.

19. Khan, S; Cao, Q; Zheng, YM; Huang, YZ; Zhu, YG. Health risks of heavy metals in contaminated soils and food crops irrigated with wastewater in Beijing, China. *Environ. Pollut* **2008**, *152*, 686–692.

20. Miller, JR; Lechler, PJ; Bridge, G. Mercury contamination of alluvial sediments within the Essequibo and Mazaruni river basins, Guyana. *Water Air Soil Pollut* **2003**, *148*, 139–166.

21. Engle, MA; Gustin, MS; Lindberg, SE; Gertler, AW; Ariya, PA. The influence of ozone on atmospheric emissions of gaseous elemental mercury and relative gaseous mercury from substrates. *Atmos. Environ* **2005**, *39*, 7506–7517.

4

# HOTSPOT ANALYSIS OF SPATIAL ENVIRONMENTAL POLLUTANTS USING KERNEL DENSITY ESTIMATION AND GEOSTATISTICAL TECHNIQUES

**Yu-Pin Lin [1], Hone-Jay Chu [1,*], Chen-Fa Wu [2], Tsun-Kuo Chang [1] and Chiu-Yang Chen [3]**

[1]Department of Bioenvironmental Systems Engineering, National Taiwan University, 1, Section 4, Roosevelt Road, Da-an District, Taipei City 106, Taiwan

[2]Department of Horticulture, National Chung Hsing University, 250, Kuo Kuang Road, Taichung 402, Taiwan

[3]Department of Safety, Health and Environmental Engineering, Ming Chi University of Technology, 84, Gungjuan Road, Taishan, Taipei 24301, Taiwan

## ABSTRACT

Concentrations of four heavy metals (Cr, Cu, Ni, and Zn) were measured at 1,082 sampling sites in Changhua county of central Taiwan. A hazard zone is defined in the study as a place where the content of each heavy metal exceeds the corresponding control standard. This study examines the use of spatial analysis for identifying multiple soil pollution hotspots in the study area. In a preliminary investigation, kernel density estimation (KDE) was a technique used for hotspot analysis of soil pollution from a set of observed occurrences of hazards. In addition, the study estimates the hazardous probability of each heavy metal using geostatistical techniques such as the sequential indicator simulation (SIS) and indicator kriging (IK). Results show that there are multiple hotspots for these four heavy metals and they are strongly correlated to the locations of industrial plants and irrigation systems in the study area. Moreover, the pollution hotspots detected using the KDE are the almost same to those estimated using IK or SIS. Soil pollution hotspots and polluted sampling densities are clearly defined using the KDE approach based on contaminated point data. Furthermore, the risk of hazards is explored by these techniques such as KDE and geostatistical approaches and the hotspot areas are captured without requiring exhaustive sampling anywhere.

## KEYWORDS

kernel density estimation (KDE); indicator Kriging (IK); sequential indicator simulation (SIS); heavy metal; soil contaminant

## 1. INTRODUCTION

Unfortunately, as a result of industrial activities, improper disposal of wastes, pollution of agricultural soils with heavy metals has become an increasingly serious problem throughout the world [1–4]. To understand the contamination risk, monitoring is a necessary and prohibitively costly process. Risk assessment at unsampled locations is of significant importance for the delineation of contamination areas [5–9]. However, the accuracy of risk estimation depends on the methodology used and various related factors.

Geostatistical analysis considers the concentration of a potentially hazard in soil as a regionalized variable in space. Geostatistics was developed as a means to describe spatial patterns of soil pollution by semivariograms and to predict the values of soil attributes at unsampled locations [10]. Geostatistical models could be used to estimate the spatial patterns of soil contaminant without measuring soil data in an entire area. The degree of contamination and hotspot areas for soils may vary with the methods used. For delineating hazardous areas, indicator kriging (IK) determines the spatial probability distribution of soil pollution in fields [6,11–16]. IK provides a non-parametric distribution estimated at an unsampled location directly using fixed thresholds and qualifies the spatial patterns of a hazardous risk. Moreover, stochastic simulation methods such as sequential indicator simulation (SIS), have been recently proposed to overcome the inherent limitations of IK [17–19]. The stochastic simulation method is based on a probabilistic model, and does not require any assumption for the shape of the conditional distribution and the systematically adds a stochastic noise component into the kriging model. Simulation with multiple realizations offers significant improvements over kriging techniques at sites with high data variations.

Hotspot mapping is used to help identify where soil pollution exists and comes from. Recently, Kernel density estimation (KDE) is one of the methods for analyzing the first order properties of a point event distribution [20–22], in part because it is easy to understand and implement. KDE has been widely used for hotspot analysis and detection. The objective of KDE is to produce a smooth density surface of point events over space by computing event intensity as density estimation [22–24]. Moreover, Schnabel and Tietje [23] applied the KDE method to spatially distributed heavy metal soil data and compared it with ordinary kriging. The results represent the interdependence between various heavy metal concentrations and additional site characteristics. Furthermore, the method could be a valuable supplement for the geostatistical uncertainty assessments.

The purpose of this study was to propose alternative approaches in searching for pollutant hotspots. The primary objective of the present work was to investigate proposals for delineating soil pollutant hazards. First, KDE identifys the hotspots of soil pollutions based on the hazardous metal sampling data. Then, IK and SIS generate a hazard probability map based on the samples for management. A study case from a field survey is provided to estimate the probability maps for hazard delineation. It is expected that results can give references for identification of hazardous areas.

# 2. METHODS AND MATERIALS

Kernel density estimation is used to identify the location, spatial extent and intensity of soil pollution hotspots. Moreover, the spatial patterns of hazardous probability for heavy metals are estimated using geostatistical methods. The three methods are used for visualization of hotspots of soil pollutions in the case study. Study area and sampling of heavy metals will be discussed in the following sections.

## 2.1. Study Area and Soil Sampling

The study area is in Changhua County, which is a critical agricultural region in Taiwan. Changhua city is in the east area and Lugang town lies to the west. Approximately 106 industrial plants are clustered in study area. Most industrial plants in the study area involve metalwork, electroplating, textile and metal surface treatment industries (Figure 1). The industrial plants have been suspected of discharging wastewater into irrigation channels in this study area [8,12,25]. The data of 1,309 topsoil (0–15 cm) samples containing concentrations of Cr, Cu, Ni, and Zn were obtained by the soil heavy metal investigation project carried by Taiwan's Environmental Protection Administration (EPA), between February and August 2002. The sampling sites are shown in Figure 1.

Approximately 1 kg of soil was collected for each sample using a stainless steel spade and a plastic scoop and then stored in a plastic food bag. After air drying at room temperature, 3 g of each soil sample were disaggregated, sieved to 0.85 mm and ground to a fine 0.15 mm powder. Each 3 g milled sample was then digested for 2 h at room temperature with 7 mL $HNO_3$ and 21 mL HCl (aqua regia, 1:3) to slowly oxidize organic matter in the soil. Next, the digest was filtered and made up to 100 mL with distilled water [15,16]. The levels of heavy metals in the samples were determined by Inductively Coupled Plasma-Optical Emission Spectrometers (ICP-OES).

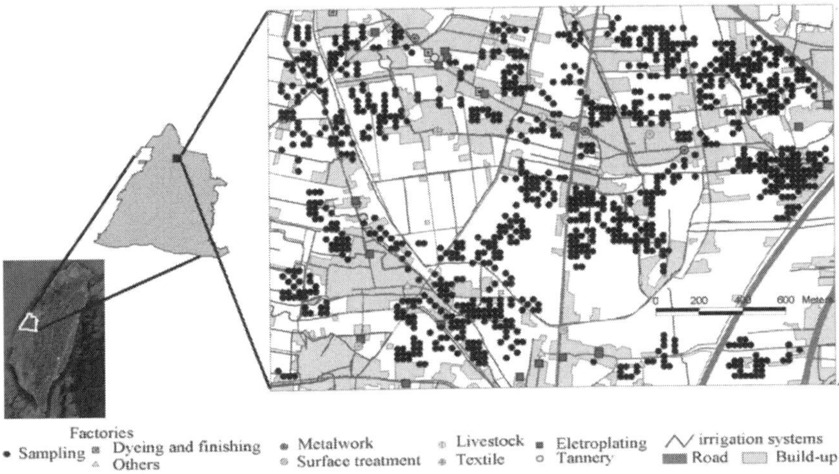


**Figure 1.** The study area and sampling sites.

## 2.2. Kernel Density Estimation (KDE)

The general form of a kernel density estimator in a 2-D space, termed KDE in the rest of this paper, is given by [22]:

$$\lambda(s) = \sum_{i=1}^{n} \frac{1}{\pi r^2} k\left(\frac{d_{is}}{r}\right) \tag{1}$$

where $\lambda$ $(s)$ is the density at location s, $r$ is the search radius (bandwidth) of the KDE, $n$ is the number of sampling points,$k$ is the weight of a point $i$ at distance $d_{is}$ to location $s$. $k$ is usually modeled as a kernel function of the ratio between $d_{is}$ and $r$.In this study, we used a kernel with a Gaussian function given by:

$$k\left(\frac{d_{is}}{r}\right) = \begin{cases} \frac{1}{\sqrt{2\pi}} \exp\left(-\frac{d_{is}^{2}}{2r^{2}}\right), & if \quad 0 < d_{is} \le r \\ 0, & otherwise \end{cases} \tag{2}$$

To identify the soil pollution hotspots, the KDE package based on ArcGIS software was used in the study.

## 2.3. Indicator Kriging (IK)

The IK estimates the probability that the concentration of a pollutant exceeds a specific control value at a given location [8,17]. The data ($z(s)$) are transformed into an indicator as follows:

$$I\ (s, z_c) = \begin{cases} 1, & if \quad z(s) \le z_c \\ 0, & otherwise \end{cases} \tag{3}$$

If the concentration of heavy metal [ $z(s)$ ] exceeds $z_c$ then the indicator is 0, otherwise it is 1 [11]. The expected value of $I(s; z_c \mid (n))$, conditional on $n$ surrounding data, can be expressed as:

$$E[I(s; z_c \mid (n))] = prob[z(s) \le z_c \mid (n)] \tag{4}$$

The hazardous probability that exceeds $z_c$ can be expressed as:

$$prob[z(s) > z_c \mid (n)] = 1 - prob[z(s) \le z_c \mid (n)] \tag{5}$$

This ordinary indicator kriging estimator is:

$$prob[z(s_0) \le z_c \mid (n)] = \sum_{\alpha=1}^{n} \lambda_\alpha I(s_\alpha; z_c) \tag{6}$$

where $I\ (s_\alpha;z_c)$ represents the indicator values at $x_\alpha$ ; $\alpha = 1,\cdots,n$ ; $\lambda_\alpha$ is the kriging weight of ($I\ (s_\alpha;z_c)$) determined by solving the following kriging system:

$$\sum_{\beta=1}^{n} \lambda_{\beta}\gamma_{i}(s_{\alpha} - s_{\beta}; z_{c}) + \mu = \gamma_{i}(s_{\alpha} - s_{0}; z_{c}) \quad (7)$$

$$\sum_{\beta=1}^{n} \lambda_{\beta} = 1 \quad (8)$$

where $\mu$ is the Lagrange multiplier; $\gamma_i$ $(s_\alpha$ − $;s_\beta;$ $z_c)$ is the indicator variogram between indicator variables at the $\alpha^{th}$ and $\beta^{th}$sampling points; $\gamma_i$ $(s_\alpha$ − $;s_0;$ $z_c)$ is the variogram between the indicator variables, *i.e.*, the $\alpha^{th}$ sampling point and $s_0$; $\alpha$ =1,···,$n$.

## 2.4. Sequential Indicator Simulation (SIS)

In sequential indicator simulation, modeling of the *N*-point conditional cumulative distribution function (ccdf) is a sequence of *N* univariate ccdfs at each grid cell along a random path [25,26]. The SIS requires the following steps [17,25,26]:

- Define a random path that visits each location of the domain once, in which all nodes $\{s_i,\ i = 1,N\}$ ··· discretizing the interest domain. A random visiting sequence ensures that no spatial continuity artifact is introduced into the simulation by a specific path visiting *N* nodes.

- At the first visited nodes $(s_1)$:

    - Model, using either a parametric or nonparametric approach, the local ccdf of $z(s_1)$ conditional on $n$ original data $\{z(s_\alpha),\alpha =1,\cdots,n\}$:

$$F_Z(s_1; z_1|(n)) = prob\{z(s_1) \le z_1|(n)\} \quad (9)$$

    - Generate, via the Monte Carlo drawing relation, a simulated value $z^{(l)}$ $(s_1)$ from this ccdf $F_Z$ $(s_1$ ; $z_1$ | $(n))$, and add it to the conditioning data set, now of dimension $n$ +1, to be used for all subsequent local ccdf determinations.

- At the $i^{th}$ node $s_i$ along the random path:
  - Model the local ccdf of $z(s_i)$ conditional on $n$ original data and the $i$ −1 near previously simulated values:

$$\{z^{(l)}(s_j), j = 1,\cdots,i-1\} : F_Z(s_i; z_i|(n+i-1)) = prob\{z(s_i) \le z_i|(n+i-1)\} \quad (10)$$

  - Generate a simulated value $z^{(l)}(s_i)$ from this ccdf, and add it to the conditioning data set, now of dimension $n + i$.
- Repeat step 3 until all $N$ nodes along the random path are visited.

The probability of soil heavy metal at $s$ exceeding the control standard ($z_c$) can be denoted by $prob[z(s) > z_c]$ [18,19]:

$$prob[z(s) > z_c] = n(s)/1000 \quad (11)$$

where $n(s)$ is the number of realizations if $z(s)$ is higher than the control standard in the 1000 SIS realizations.

# 3. RESULTS AND DISCUSSION

## 3.1. Basic Statistics

Table 1 summarizes the descriptive statistics of the investigated four heavy metals (Cr, Cu, Ni, and Zn) from the original 1,082 samples. In Taiwan, the pollution control standards (maximum allowable concentrations) for the investigated heavy metals are as follows (in mg/kg): Cr 250, Cu 200, Ni: 200 and Zn 600. Table 1 lists 286 samples for Cr, 395 samples for Cu, 622 samples for Ni, and 336 samples for Zn over the control standards. Moreover, the high variability of the pollutant concentrations at various heavy metals requires a detailed evaluation and interpretation. The application of various methods is an efficient tool in achieving better understanding of the hazardous state of the soil pollution. It seems recommendable to combine various approaches instead of relying only on one of them to gain better information of the pollutions, such as the KDE, geostatistical methods (*i.e.*, the IK and SIS).

**Table 1.** Descriptive statistics of heavy metals for 1,082 samples.

| | Min (mg/kg) | Median (mg/kg) | Max (mg/kg) | Average (mg/kg) | SD (mg/kg) | Control standards (mg/kg) | Number of observances over control standards |
|---|---|---|---|---|---|---|---|
| Cr | 22.6 | 119.0 | 3,070.0 | 194.0 | 212.5 | 250 | 286 |
| Cu | 11.0 | 116.0 | 3,810.0 | 194.7 | 222.7 | 200 | 395 |
| Ni | 21.3 | 189.2 | 4,020.0 | 271.3 | 259.0 | 200 | 622 |
| Zn | 60.5 | 337.0 | 7,850.0 | 526.4 | 549.6 | 600 | 336 |

Min: minimum; Max: maximum; SD: standard deviation.

## 3.2. Point Pattern Analysis Using Kernel Density Estimation (KDE)

Figure 2 shows the hotspot patterns of soil pollutants on the kernel density map. The Kernel Density Estimation (KDE) transforms a dot pattern into a continuous surface, providing a more useful representation of soil pollution distributions, allowing for easier detection of possible pollution hotspots [24]. Results show that the hotspots associated with the four heavy metals in the study area are often multiple. We found that soil pollution hotspots were more clearly defined using KDE, probably because of the clustered distribution of soil pollution occurrences. The spatial patterns also reveal Cr hotspots near industrial plants and irrigation systems in the study area. The areas with Cu hotspots are in the central and eastern parts of the study area in the vicinity of the industrial plants and irrigation systems. Hotspots of Ni are particularly highly distributed throughout the northeast part of the studied area. The areas with Zn hotspots are close to the industrial plants and irrigation systems in the northwest. However, there are potentially contaminated sites that are hidden, especially in the areas where the history of land use is complicated and the sources of imported soils are usually unknown [27]. The maps show that the area with high susceptibility of pollution is along the industrial plants and the irrigation systems. The KDE results match the previous studies showing that the distributions of background heavy metals and pollution sources correlated with industrial plants and irrigation channels [15]. These industrial plants are suspected of discharging wastewater into irrigation systems in the study area [12,15,16,25].

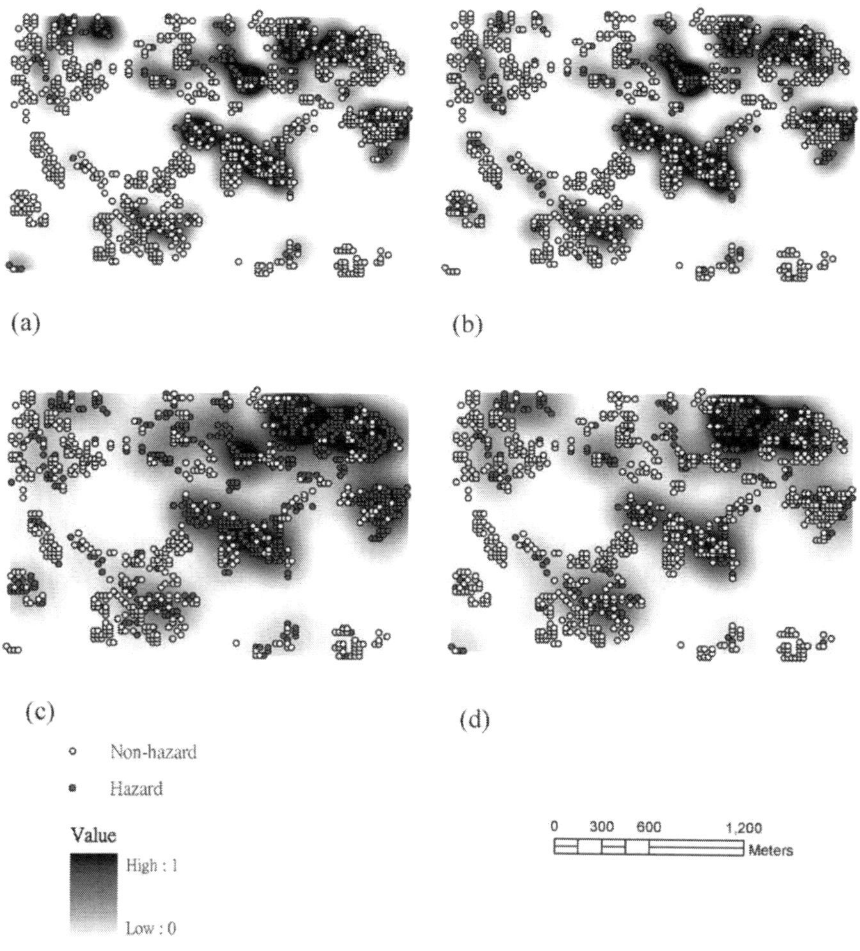


**Figure 2.** The kernel density maps (stretched to min–max range) of (a) Cr (b) Cu (c) Ni (d) Zn.

## 3.3. Sampling Density and Spatial Interpolation of Probability Exceeding Control Standards Using the IK and SIS Approaches

The hazardous probability that heavy metal concentrations exceed control standards at any of the unsampled sites is determined by geostatistical methods (*i.e.*, IK and SIS). Moreover, the spatial distribution of hazardous probability can be characterized by an indicator variogram. The variance is estimated as a function of a variogram model, where the variogram is calculated using the relative locations of the samples. Table 2 lists the parameters of indicator variograms for the four heavy metals. In the indicator variograms, the fitted ranges, the nugget effects and the sills are 120–249 m, 0.0206–0.0251 and 0.187–0.235 (Table 2), respectively. The results show that the sill value for Ni is the largest one. Based on the cases, the higher sill corresponds to greater variability in the probability map. Spatial structure analysis has been identified to be a useful tool in illustrating the spatial patterns of variables, and a necessary basis for a number of other spatial analysis procedures, such as kriging analysis [28]. Furthermore, SIS realizations are performed based on the indicator variogram models for the 25th, 50th, and 75th percentiles of the sample distribution (Table 3) of original samples for Cr, Cu, Ni, and Zn in the study area.

**Table 2.** Indicator variogram models for heavy metals.

| | Threshold (mg/kg) | Model | $C_0$ | $C_0$+C | R (m) | RSS | $r^2$ |
|---|---|---|---|---|---|---|---|
| Cr | 250 | Exp. | 0.0237 | 0.1874 | 120 | 2.52E-04 | 0.859 |
| Cu | 200 | Exp. | 0.0251 | 0.2202 | 135 | 3.08E-04 | 0.904 |
| Ni | 200 | Exp. | 0.0206 | 0.2352 | 249 | 3.39E-03 | 0.723 |
| Zn | 600 | Exp. | 0.0221 | 0.2042 | 147 | 6.05E-04 | 0.808 |

Exp.: Exponential model; $C_0$: Nugget; $C_0$+C: Sill; R: Range; RSS: Residual Sum of Squares; $r^2$: Coefficient of determination

Figures 3 and 4 show the probability maps for sites where Cr, Cu, Ni, and Zn exceed the control standards based on the IK and 1000 SIS realizations (by Equations 5 and 11).

**Table 3.** Indicator variogram models for the 25th, 50th, and 75th percentiles of heavy metals in 1,082 samples.

| Heavy metal | | Model | Parameters | | | RSS | $r^2$ |
|---|---|---|---|---|---|---|---|
| | | | $C_0$ | $C_0$+C | R (m) | | |
| Cr | 25% | Exp. | 0.020 | 0.184 | 216 | 1.730E-03 | 0.722 |
| | 50% | Exp. | 0.026 | 0.247 | 171 | 1.202E-03 | 0.807 |
| | 75% | Exp. | 0.025 | 0.190 | 120 | 2.075E-04 | 0.852 |
| Cu | 25% | Exp. | 0.017 | 0.184 | 240 | 2.008E-03 | 0.737 |
| | 50% | Exp. | 0.025 | 0.247 | 186 | 7.016E-04 | 0.899 |
| | 75% | Exp. | 0.024 | 0.190 | 108 | 5.293E-04 | 0.663 |
| Ni | 25% | Exp. | 0.015 | 0.179 | 222 | 2.614E-03 | 0.634 |
| | 50% | Exp. | 0.022 | 0.237 | 228 | 3.608E-03 | 0.671 |
| | 75% | Exp. | 0.018 | 0.183 | 159 | 5.723E-04 | 0.805 |
| Zn | 25% | Exp. | 0.024 | 0.190 | 222 | 1.464E-03 | 0.768 |
| | 50% | Exp. | 0.028 | 0.250 | 171 | 3.795E-04 | 0.936 |
| | 75% | Exp. | 0.021 | 0.189 | 144 | 8.077E-03 | 0.710 |

Exp.: Exponential model; $C_0$: Nugget; $C_0$+C: Sill; R: Range; RSS: Residual Sum of Squares; $r^2$: Coefficient of determination.

The results demonstrate that the hotspots of hazard probability for Cr and Cu are similar. The spatial patterns of hazard probability also reveal hotspots of Cr near industrial plants and irrigation systems in the study area. The Cu hotspots are located in the central and east-northern parts of the study area in the vicinity of industrial plants and irrigation systems. The hotspots of Ni are distributed throughout the study area, except for the south-western part; and the areas with high concentrations of Zn are close to industrial plants and irrigation systems in the north-western part. Furthermore, all probability maps show that the multiple hotspots of hazard probability are close to industrial plants and irrigation systems in the study area.

Table 4 shows that polluted sampling density value is subjected to SIS probability exceeding regulatory thresholds with given critical probabilities ($p_c$ = 0.9, 0.8, 0.7 and 0.6) in the Cr, Cu, Ni, and Zn content of soil. Results show polluted sampling density increases as the critical probability increases. The polluted sampling density could be detected using KDE when delineating contaminations based on the original samples. In the study area, polluted sampling density range from 0.00023 to 0.00036 (L/m$^2$) However, density values for heavy metal Ni are the lowest among the four heavy metals. For long-term pollution monitoring, the Ni pollution sampling points could be increased primarily. Based on these results, the KDE method is an effective approach to make sure of sampling density in delineating heavy metal pollutions for further monitoring.

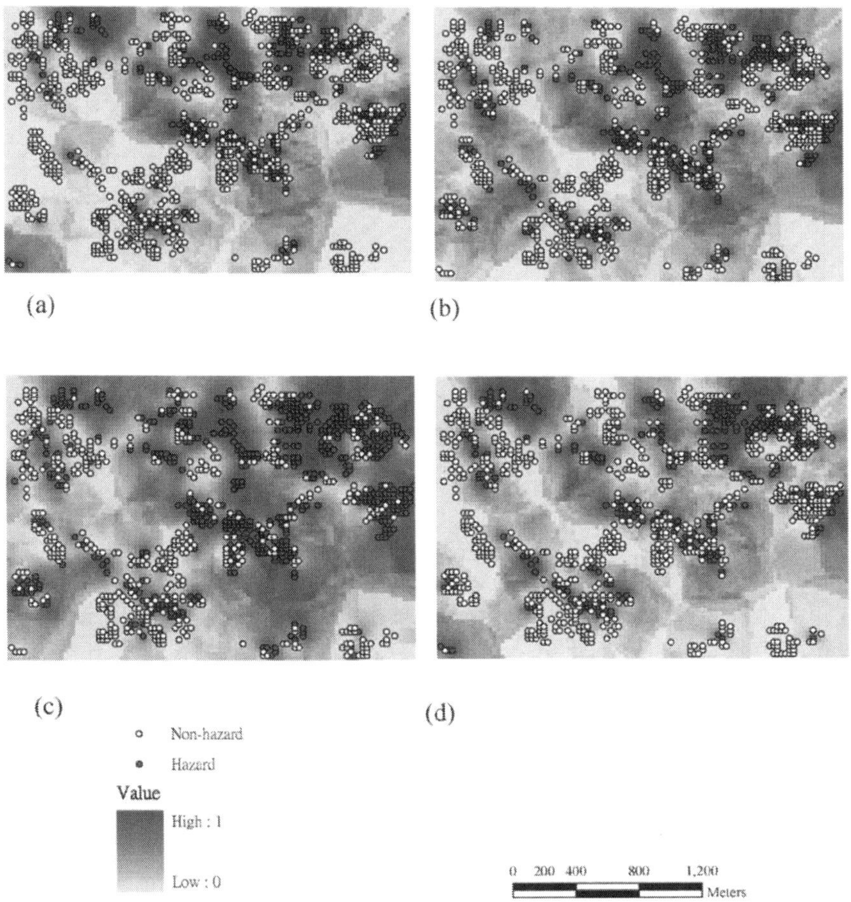


**Figure 3.** The probability maps of (a) Cr (b) Cu (c) Ni (d) Zn using indicator kriging based on 1,082 samples.

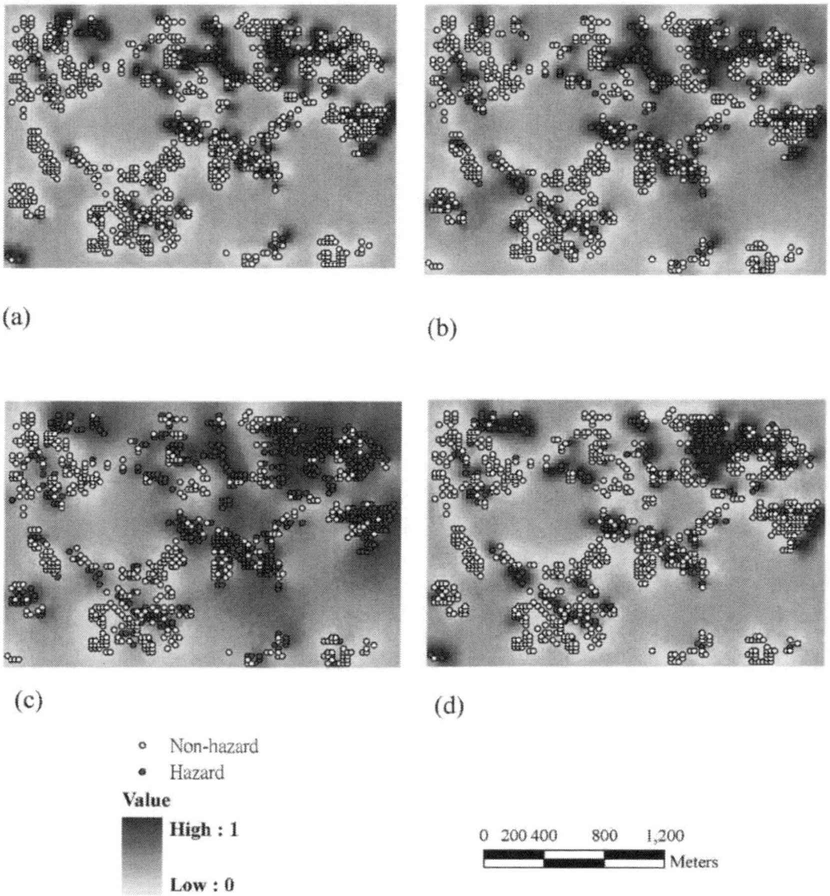


**Figure 4.** The probability maps of (a) Cr (b) Cu (c) Ni (d) Zn in 1000 realizations using sequential indicator simulation based on 1,082 samples.

## 3.4. Comparisons of Hotspot Visualizations by Various Methods

These techniques such as KDE, IK and SIS are commonly used in exploratory spatial analyses and pattern resolution for soil pollution visualization of heavy metals. All three visualization methods that we used to explore the soil pollution intensity patterns showed similar results (Figures 2–4) near factories and irrigation systems. These methods showed generally consistent results, but differences existed. KDE is an efficient means of detecting soil pollution hotspots based on point data. Results show the pollution hotspots are consistent in the other two approaches (*i.e.*, IK and SIS). The KDE results also show multiple hotspots in the study area and may under-emphasized areas with heavy metal pollution. However, the hotspots when determined based on KDE are more conservative than the ones estimated by IK and SIS. Results imply that the KDE multiple hot spots may be under-emphasized heavy metal pollution zones in the study area (Figure 2). Estimation results indicate that the hazard probability patterns estimated by the SIS are less fragmented than those estimated by IK (Figures 3 and 4). SIS takes into account not only the spatial variation of observed data at sampled locations but also the variation in estimations at unsampled locations which kriging estimation ignores those factors [18,25]. In addition, the simulation approach modifies the failure of IK to reproduce clusters of large concentrations above the tolerable maximum [18]. Simulation generates equally likely sets of values for a variable, which are consistent with available *in-situ*measurements. This usually implies that the simulated values have the same mean and variogram as the original data; they may also have to coincide with the original data at sampling points [29]. Furthermore, the local uncertainty information obtained by the IK is not sufficient to quantify the uncertainty at several locations simultaneously. Future work could assess multi-location uncertainties using SIS for the delineation of soil pollution. In addition, several investigators have published evidence of dynamics in environmental management [30–33]. In the future study, the temporal analysis of pollutant concentrations could be further explored.

# 4. CONCLUSIONS

This study utilizes KDE and geostatistical techniques with 1,082 samples to delineate hazardous zones and quantify the risk of multiple pollutants in a contaminated area. Various methodologies show generally consistent results, but differences exist. The results demonstrate that KDE is an alternative means of determining hazardous hotspots of soil pollutants only using hazardous point data in the preliminary investigation. The polluted sampling density could be detected by using KDE with SIS delineation. Moreover, the geostatistical models are approaches for identifying the risk of hazard delineation and are highly promising for use in evaluating the susceptibility of heavy metals without surveying soil concentrations over an entire study area. All proposed methods can be extended to show that soil pollution is closely related to pollution sources such as industrial factories and the irrigation system in the study area. According to the spatial maps, model assessment of soil pollution hotspots enables remediation planners to help identify hazardous pollution areas. Integrating KDE and geostatistical methods, the KDE method is an effective approach to determine sampling density when delineating heavy metal pollutions by geostatistical methods. The information of spatial sampling density and hotspot pattern could be useful for long-term monitoring and assessment.

**Table 4.** Polluted sampling density value based on SIS probability criteria.

| | Critical probability( $p_c$ ) | Number of grid which value is over $p_c$ | Density value (L/m$^2$) | |
|---|---|---|---|---|
| | | | Mean | Range |
| Cr | 0.6 | 591 | 0.00028 | 0.00066 |
| | 0.7 | 467 | 0.00031 | 0.00066 |
| | 0.8 | 373 | 0.00033 | 0.00066 |
| | 0.9 | 310 | 0.00034 | 0.00066 |
| Cu | 0.6 | 851 | 0.00029 | 0.00082 |
| | 0.7 | 643 | 0.00032 | 0.00081 |
| | 0.8 | 505 | 0.00034 | 0.00080 |
| | 0.9 | 403 | 0.00036 | 0.00080 |
| Ni | 0.6 | 2,157 | 0.00023 | 0.00071 |
| | 0.7 | 1,554 | 0.00027 | 0.00070 |
| | 0.8 | 1,099 | 0.00030 | 0.00067 |
| | 0.9 | 773 | 0.00032 | 0.00067 |
| Zn | 0.6 | 709 | 0.00028 | 0.00079 |
| | 0.7 | 560 | 0.00030 | 0.00079 |
| | 0.8 | 453 | 0.00032 | 0.00079 |
| | 0.9 | 379 | 0.00033 | 0.00079 |

## ACKNOWLEDGEMENTS

The authors wish to thank the anonymous reviewers for their valuable comments and suggestions. The authors would like to thank the National Science Council of the Republic of China, Taiwan, for financially supporting this research under Contract NSC-97-2628-H-002-026-MY3, NSC 89-2621-B- 002-004 and NSC 87-2621-P-002-012.

## REFERENCES AND NOTES

1. Yu, H; Wang, JL; Fang, W; Yuan, JG; Yang, ZY. Cadmium accumulation in different rice cultivars and screening for pollution-safe cultivars of rice. *Sci. Total E* **2006**, *370*, 302–309.
2. Hossain, MF. Arsenic contamination in Bangladesh—an overview. *Agr. Eco. Env* **2006**, *113*, 1–16.
3. Rehman, W; Zeb, A; Noor, N; Nawaz, M. Heavy metal pollution assessment in various industries of Pakistan. *Envir. Geol* **2008**, *55*, 353–358.
4. Shomar, B. Sources and build up of Zn, Cd, Cr and Pb in the sludge of Gaza. *Env. Mon. Ass* **2009**, *155*, 51–62.
5. Goovaerts, P. Geostatistical modelling of uncertainty in soil science. *Geoderma* **2001**, *103*, 3–26.
6. van Meirvenne, M; Goovaerts, P. Evaluating the probability of exceeding a site-specific soil cadmium contamination threshold. *Geoderma* **2001**, *102*, 75–100.
7. Amini, M. Mapping risk of cadmium and lead contamination to human health in soils of Central Iran. *Sci. Total E* **2005**,*347*, 64–77.

8. Lin, YP. Multivariate geostatistical methods to identify and map spatial variations of soil heavy metals. *Envir. Geol* **2002**, *42*, 1–10.

9. Hassan, MM; Atkins, PJ. Arsenic risk mapping in Bangladesh: a simulation technique of cokriging estimation from regional count data. *J. Environ. Sci. Health A* **2007**, *42*, 1719–1728.

10. Goovaerts, P. Geostatistics in soil science: state-of-the-art and perspectives. *Geoderma* **1999**, *89*, 1–45.

11. Goovaerts, P. Soares, A, Gómez-Hernández, J, Froidevaux, R, Eds.; Kriging *vs.* stochastic simulation for risk analysis in soil contamination. In *geoENV I—Geostatistics for Environmental Applications*; Kluwer Academic Publishers: Dordrecht, The Netherlands, 1997; pp. 247–258.

12. Lin, YP; Chang, TK; Shih, CW; Tseng, CH. Factorial and indicator kriging methods using a geographic information system to delineate spatial variation and pollution sources of soil heavy metals. *Envir. Geol* **2002**, *42*, 900–909.

13. Brus, DJ; de Gruijter, JJ; Walvoort, DJJ; de Vries, F; Bronswijk, JJB; Romkens, PFAM; de Vries, W. Mapping the probability of exceeding critical thresholds for cadmium concentrations in soils in the Netherlands. *J. Envir. Q* **2002**,*31*, 1875–1884.

14. Huang, SS; Liao, QL; Hua, M; Wu, XM; Bi, KS; Yan, CY; Chen, B; Zhang, XY. Survey of heavy metal pollution and assessment of agricultural soil in Yangzhong district, Jiangsu Province, China. *Chemosphere* **2007**, *67*, 2148–2155.

15. Lin, YP; Cheng, BY; Shyu, GS; Chang, TK. Combining a finite mixture distribution model with indicator kriging to delineate and map the spatial patterns of soil heavy metal pollution in Chunghua County, central Taiwan. *Environ. Pollut* **2010**, *158*, 235–244.

16. Chu, HJ; Lin, YP; Jang, CS; Chang, TK. Delineating the hazard zone of multiple soil pollutants by multivariate indicator kriging and conditioned Latin hypercube sampling. *Geoderma* **2010**, *158*, 242–251.

17. Deutsch, CV; Journel, AG. *GSLIB, Geostatistical Software Library and User's Guide*; Oxford University Press: New York, NY, USA, 1992.

18. Juang, KW; Chen, YS; Lee, DY. Using sequential indicator simulation to assess the uncertainty of delineating heavy-metal contaminated soils. *Environ. Pollut* **2004**, *127*, 229–238.

19. Zhao, YC; Shi, XZ; Yu, DS; Wang, HJ; Sun, WX. Uncertainty assessment of spatial patterns of soil organic carbon density using sequential indicator simulation, a case study of Hebei province, China. *Chemosphere* **2005**, *59*, 1527–1535.

20. Bailey, TC; Gatrell, AC. *Interactive Spatial Data Analysis*; Longman: Essex, UK, 1995.

21. Silverman, BW. *Density Estimation for Statistics and Data Analysis*; Chapman Hall: London, UK, 1986.

22. Xie, ZX; Yan, J. Kernel Density Estimation of traffic accidents in a network space. *Comput. Environ. Urban* **2008**, *32*, 396–406.

23. Schnabel, U; Tietje, O. Explorative data analysis of heavy metal contaminated soil using multidimensional spatial regression. *Envir. Geol* **2003**, *44*, 893–904.

24. Serra-Sogas, N; O'Hara, PD; Canessa, R; Keller, P; Pelot, R. Visualization of spatial patterns and temporal trends for aerial surveillance of illegal oil discharges in western Canadian marine waters. *Mar. Pollut. Bull* **2008**, *56*, 825–833.

25. Lin, YP; Chu, HJ; Hwang, YL; Chen, BY; Chang, TK. Modeling spatial uncertainty of heavy metals content in soil by conditional

latin hypercube sampling and geostatistical simulation. *Environ. Earth Sciences* **2010**. (in press).

26. Kyriakidis, PC. Hunsaker, CT, Goodchild, MF, Friedl, MA, Case, TJ, Eds.; Geostatistical models of uncertainty for spatial data. In *Spatial Uncertainty in Ecology: Implications for Remote Sensing and GIS Applications*; Springer-Verlag: New York, NY, USA, 2001; pp. 175–213.

27. Zhang, C; Tang, Y; Luo, L; Xu, W. Outlier identification and visualization for Pb concentrations in urban soils and its implications for identification of potential contaminated land. *Environ. Pollut* **2009**, *157*, 3083–3090.

28. Wang, XJ; Qi, F. The effects of sampling design on spatial structure analysis of contaminated soil. *Sci. Total Environ* **1998**, *224*, 29–41.

29. Lin, YP; Chang, TK. Geostatistical simulation and estimation of the spatial variability of soil zinc. *J. Environ. Sci. Health A* **2000**, *35*, 327–347.

30. Chang, LC; Chu, HJ; Hsiao, CT. Optimal planning of a dynamic pump-treat-inject groundwater remediation system. *J. Hydrol* **2007**, *342*, 295–304.

31. Telesca, L; Caggiano, R; Lapenna, V; Lovallo, M; Trippetta, S; Macchiato, M. The Fisher information measure and Shannon entropy for particulate matter measurements. *Phys. Stat. Mech. Appl* **2008**, *387*, 4387–4392.

32. Telesca, L; Caggiano, R; Lapenna, V; Lovallo, M; Trippetta, S; Macchiato, M. Analysis of dynamics in Cd, Fe and Pb particulate matter by using the Fisher-Shannon method. *Water Air Soil Pollut* **2009**, *201*, 33–41.

33. Yu, HL; Chu, HJ. Understanding space-time patterns of groundwater systems by empirical orthogonal functions: a case study in the Choshui River Alluvial Fan, Taiwan. *J. Hydrol* 2010, *381*, 239–247.

5

# ASSESSMENT OF WATER QUALITY IN A SUBTROPICAL ALPINE LAKE USING MULTIVARIATE STATISTICAL TECHNIQUES AND GEOSTATISTICAL MAPPING: A CASE STUDY

**Wen-Cheng Liu [1,*], Hwa-Lung Yu [2] and Chung-En Chung [1]**

[1]Department of Civil Disaster Prevention Engineering, National United University, Miao-Li, 36003, Taiwan

[2]Department of Bioenvironmental Systems Engineering, National Taiwan University, Taipei, 10617, Taiwan

## ABSTRACT

Concerns about the water quality in Yuan-Yang Lake (YYL), a shallow, subtropical alpine lake located in north-central Taiwan, has been rapidly increasing recently due to the natural and anthropogenic pollution. In order to understand the underlying physical and chemical processes as well as their associated spatial distribution in YYL, this study analyzes fourteen physico-chemical water quality parameters recorded at the eight sampling stations during 2008–2010 by using multivariate statistical techniques and a geostatistical method. Hierarchical clustering analysis (CA) is first applied to distinguish the three general water quality patterns among the stations, followed by the use of principle component analysis (PCA) and factor analysis (FA) to extract and recognize the major underlying factors contributing to the variations among the water quality measures. The spatial distribution of the identified major contributing factors is obtained by using a kriging method. Results show that four principal components *i.e.*, nitrogen nutrients, meteorological factor, turbidity and nitrate factors, account for 65.52% of the total variance among the water quality parameters. The spatial distribution of principal components further confirms that nitrogen sources constitute an important pollutant contribution in the YYL.

## KEYWORDS

multivariate statistical technique; geostatistical mapping; water quality; principal component analysis; cluster analysis; Yuan-Yang Lake

## 1. INTRODUCTION

Water quality is the main factor controlling healthly and diseased states in both humans and animals. Surface water quality is an essential component of the natural environment and a matter of serious concern today. The variations of water quality are essentially the combination of both anthropogenic and natural contributions. In general, the anthropogenic discharges constitute a constant source of pollution, whereas surface runoff is a seasonal phenomenon which is affected by climate within the water catchment basin [1]. Among them, because of the intensive human activities, the anthropogenic inputs from a variety of sources are commonly the primary factors affecting the water quality of most rivers, lakes, estuaries, and seas, especially for those close to highly urbanized regions.

Many investigations have been conducted on anthropogenic contaminants of ecosystems [2–4]. Because of the spatial and temporal variations in water quality conditions, a monitoring program which provides a representative and reliable estimation of the quality of surface waters is necessary. The monitoring results produce a large and complicated data matrix that is difficult to interpret to draw meaningful conclusions. Multivariate statistical techniques are powerful tools for analyzing large numbers of samples collected in surveys, classifying assemblages and assessing human impacts on water quality and ecosystem conditions.

The application of different multivariate statistical techniques, such as principal component analysis (PCA), factor analysis (FA), cluster analysis (CA), and discriminate analysis (DA), assists in the interpretation of complex data matrices for a better understanding of water quality and ecological characteristics of a study area. These techniques provide the identification of possible sources that affect water environmental systems and offer a valuable tool for reliable management of water resources as well as rapid solution for pollution issues [5–7]. Multivariate statistical techniques have been widely adopted to analyze and evaluate surface and freshwater water quality, and are useful to verify temporal and spatial variations caused by natural and anthropogenic factors linked to seasonality [8,9].

Geostatistical mapping is based on field observations. Because field surveys are limited by the cost of sampling, only sparse observation data are generally available. Geostatistical mapping or further analysis requires the assessment of exhaustive attribution values for an entire study area. Geostatistical mapping techniques have been widely applied to different fields including water quality in bays [10] watersheds [11], soil properties [12], precipitation [13], river discharges [14], air pollution [15], and so on. To the best of our knowledge, geostatistical mapping has not been adopted for studying water quality data in lakes.

The objective of the present study was to analyze 14 physico-chemical water quality parameters in water samples collected on monthly basis from 2008 to 2010 in a subtropical alpine lake (Yuan-Yang Lake) in Taiwan. The data matrix obtained from field measurement was subjected to the CA, PCA, and FA techniques, as well as geostatistical mapping to evaluate information about the similarities between sampling stations and to ascertain the important contributions of nutrient sources among water quality parameters in the alpine lake.

# 2. MATERIALS AND METHODS

## 2.1. Study Site and Sample Collection

The Long-Term Ecological Research (LTER) program is one of the core projects of the Global Change and Terrestrial Ecosystem program (GCTE), which is under the umbrella of the International Geosphere-Biosphere Program (IGBP). An understanding of ecological processes and of mechanisms leading to ecologically tragic events is particularly important for the sustainability of Taiwan Island. To meet such a requirement, the LTER project was initiated in 1992 on the island. Yuan-Yang Lake (YYL) is one of the six LTER sites and the only site associated with a mountain lake ecosystem in Taiwan. YYL, a small (3.6 ha) and shallow (4.5 m maximum depth) lake in a mountainous catchment 1,730 m above sea level, is located in the northeastern region of Taiwan (24°35′ N, 121°24′E) (Figure 1). The lake and surrounding catchment (374 ha) were designated as a long-term ecological study site by the Taiwan National Science Council in 1992 and joined the Global Lake Ecological Observatory Network (GLEON) in 2004. The lake is an important site for studying physical characteristics, water quality, and ecosystems. Recently, the lake has been subject to pollution sources from recreational activities, therefore the investigation of water quality is urgent and necessary.

The steep watersheds are dominated by pristine Taiwan false cypress [*Chamaecyparis obtusa* Sieb. & Zucc. var. formosana (Hayata) Rehder] forest. The average annual temperature is approximately 13 °C (monthly average ranges from −5 to 15 °C) and the annual precipitation is more than 4,000 mm. YYL is subject to three to seven typhoons in summer and autumn each year, during which more than 1,700 mm of precipitation may fall on the lake.

The sampling network including eight measured stations was designed to cover a wide range of key locations accounting for inflow and outflow (Figure 1). Stations 1 and 2 are located at shallow area which is a swamp (shallow) zone. Stations 3 to 8 are located at the middle and deep zones. Station 4 is near by water inflow site, while station 5 is close to the site of lake water outflow.

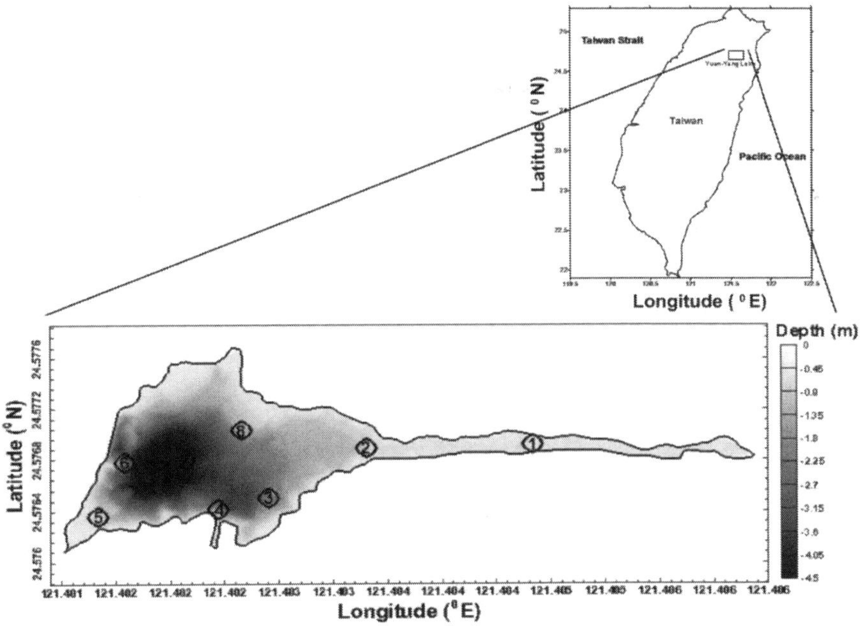


**Figure 1.** Location of Yuan-Yang Lake (YYL) in Taiwan and eight measurement stations in YYL.

Water temperatures were measured through the water column at 0.5 m increments using a thermistor chain (Templine, Apprise Technologies, Inc. Duluth, MN, USA). Wind speed was measured 1 m above the lake by an anemometer (model 03001, R.M. Young, Traverse, MI, USA). Precipitation, air temperature and downwelling photosynthetically active radiation (PAR) were measured at a land-based meteorological station approximately 1 km away from the lake. Variation in water levels was measured using a submersible pressure transmitter [PS 9800(1), Instrumentation Northwest, Kirkland, WA, USA] deployed at the lake shore (Figure 1). The attenuation of irradiance by the water column, in the 400–700 nm bands, was measured using a Licor underwater quantum flat head sensor. The outputs from the senor were stored using Licor data logger in the field, and converted to light measurements in the laboratory.

The pH, turbidity, and Secchi depth were measured *in situ*. Dissolved oxygen concentration was measured with a dissolved oxygen meter (Yellow Springs Instruments Company USA, Model 550A). The water samples, collected using an open water grab sampler equipped with a sample pull-ring that allowed for sampling at different water depths, were analyzed and measured in laboratory to obtain total suspended solids (TSS), nutrients (nitrate nitrogen, ammonium nitrogen, total nitrogen, and total phosphorus), and chlorophyll a concentrations. Chlorophyll a was measured by filtering with 600 $cm^3$ samples through a glass fiber filter. The filter paper itself was used for the analysis. The filtering was group up 90% acetone solution and fluorometer is used to read the light transmission, which in turn was used to calculate the concentration of chlorophyll a. TSS and nutrients, concentration was analyzed using the US EPA standard method 160.1 [16].

## 2.2. Cluster Analysis

CA is an unsupervised pattern recognition method that divides a large amount of cases into smaller groups or clusters based on the characteristics they process. The resulting clusters of objects should exhibit high internal (within cluster) homogeneity and high external (between clusters) heterogeneity. Hierarchical CA is the most common approach, which starts with each case in a separate cluster and joints the clusters together step by step until only one cluster remains and is typically illustrated by a dendrogram (tree diagram). The dendrogram provides a visual summary of the clustering process, presenting a picture of the groups and their proximity, with a dramatic reduction in dimensionality of original data. The Euclidean distance usually provides the similarity between two samples and a distance can be represented by the difference between analytical values from samples. In the present study, hierarchical CA was adopted to the standardized data using Ward's method, with Euclidean distance as a measure of similarity. The Ward method applies an analysis of variance approach to assess the distances between clusters to minimize the sum of squares of any two clusters that can be formed at each step. The spatial variability of water quality in the lake was determined from hierarchical CA using the linkage distance [17–19].

## 2.3. Principal Component Analysis/Factor Analysis

Principal component analysis is a data analysis method focused on a particular collection of variables. Consider the form of the first principal component. The score for individual i on component, $c_{i1}$, uses weight $w_{11}$, ....., $w_{p1}$ in the linear combination:

$$c_{i1} = y_{i1}w_{11} + y_{i2}w_{22} + \ldots + y_{ip}w_{p1} \quad (1)$$

The linear combination is chosen so that the sum of squares of $c_1$ is as large as possible subject to the condition that $w_{11}^2 + \ldots\ldots + w_{p1}^2 = 1$. The second principal component is another linear combination of $y_j$:

$$c_{i2} = y_{i1}w_{12} + y_{i2}w_{22} + \ldots + y_{ip}w_{p2} \quad (2)$$

where the variance $c_2$ is the maximal, subject to the conditions that corr ($c_1$, $c_2$ )=0 and that $w_{12}^2 + \ldots\ldots + w_{p2}^2 = 1$. The criterion of summarizing the information in *p* variables by a few components is valuable as a means of reducing the number of variables needed in an analysis [20].

FA follows PCA. FA focuses on reducing the contribution of less significant variables to simplify even more of the data structure coming from PCA. This purpose can be implemented by rotating the axis defined by PCA based on well established rules, and constructing new variables, also called varifacrors (VFs). PCA of the normalized variables was performed to extract significant PCs and to further reduce the contribution of variables with minor significance; these PCs were subjected to varimax rotation (raw) generating VFs [21,22].

The FA can be written as:

$$y_{ji} = f_{j1}z_{i1} + f_{j2}z_{i2} + \ldots + f_{jm}z_{im} + e_{ij} \quad (3)$$

where *y* is the measured variable, *f* is the factor loading, *z* is the factor score, *e* is the residual term accounting for errors, *i* is the sample number and *m* is the total number of factors. The multivariate statistical technique calculations were implemented using STATISTICA 8 [23] and Microsoft Office Excel 2007.

## 2.4. Geostatistical Mapping

Geostatistical mapping can be defined as the analytical production of maps by using field observations, auxiliary information and a computer program that generates predictions. The isotropic semivariogram are estimated to characterize the relationship between general spatial dependence and distance among the observations. Different semivariogram models, e.g., exponential and Gaussian models, nested with nugget effects are selected separately with respect to different principle components or factor scores. The optimal parameters for semivariogram models are calculated by the weighted least squares method [24]. Despite the concerns about the spatial non-orthogonality, the cross-correlations between different principle components or factor scores are calculated [25,26]. It shows that the cross-correlations increase as the spatial lags increases; however, the maximum cross-correlations are still small and less than 0.4. This study then assumes the spatial orthogonality of the principle components as well as the factor scores. The use of simple kriging usually requires the knowledge of the underlying space/time trend of the attributes of concern. However, it is not available for the modeling of "transformed" variables in this study. In these cases, many studies use nonparametric method for the trend modeling. Therefore, in this study, ordinary kriging is used for the spatial mapping which considers a non-parametric trend as well the spatial association among the attributes concurrently. All the geostatistical analysis computations of this study were performed on SEKSGUI, which is freely and publicly available [27].

# 3. RESULTS AND DISCUSSION

The measured results of 14 physico-chemical water quality parameters at eight sampling stations from August 2008 to June 2010 in the YYL are presented in Table 1.

**Table 1.** Results of water quality parameters at eight sampling in the YYL.

| Parameter | Abbreviation | Station 1 | Station 2 | Station 3 | Station 4 | Station 5 | Station 6 | Station 7 | Station 8 |
|---|---|---|---|---|---|---|---|---|---|
| Temperature (°C) | Temp | 12.4 ± 2.88 | 13.63 ± 3.80 | 14.30 ± 3.41 | 14.41 ± 3.66 | 14.67 ± 3.62 | 13.86 ± 3.24 | 13.83 ± 3.49 | 14.47 ± 3.75 |
| Dissolved Oxygen (mg/L) | DO | 5.82 ± 0.89 | 6.49 ± 0.92 | 6.85 ± 0.75 | 6.79 ± 1.08 | 6.57 ± 1.26 | 6.11 ± 1.40 | 6.01 ± 1.30 | 6.78 ± 0.82 |
| Secchi Depth (m) | SD | 0.65 ± 0.12 | 0.86 ± 0.14 | 1.79 ± 0.39 | 1.69 ± 0.40 | 1.79 ± 0.44 | 1.95 ± 0.41 | 1.92 ± 0.39 | 1.84 ± 0.36 |
| Total Phosphorus (mg/L) | TP | 0.011 ± 0.005 | 0.014 ± 0.008 | 0.012 ± 0.006 | 0.011 ± 0.006 | 0.009 ± 0.004 | 0.009 ± 0.004 | 0.010 ± 0.004 | 0.009 ± 0.003 |
| Total Nitrogen (mg/L) | TN | 0.528 ± 0.169 | 0.544 ± 0.219 | 0.452 ± 0.196 | 0.427 ± 0.115 | 0.432 ± 0.144 | 0.454 ± 0.184 | 0.448 ± 0.166 | 0.422 ± 0.169 |
| Ammonium Nitrogen (mg/L) | $NH_4$-N | 0.080 ± 0.112 | 0.078 ± 0.057 | 0.074 ± 0.039 | 0.051 ± 0.037 | 0.055 ± 0.031 | 0.097 ± 0.085 | 0.100 ± 0.102 | 0.077 ± 0.089 |
| Nitrate Nitrogen (mg/L) | $NO_3$-N | 0.111 ± 0.053 | 0.071 ± 0.038 | 0.083 ± 0.045 | 0.092 ± 0.042 | 0.091 ± 0.044 | 0.097 ± 0.044 | 0.095 ± 0.041 | 0.097 ± 0.045 |
| Total Suspended Solids (mg/L) | TSS | 5.38 ± 4.02 | 5.87 ± 3.88 | 3.79 ± 2.73 | 3.19 ± 1.95 | 4.18 ± 2.84 | 3.44 ± 3.07 | 3.90 ± 3.73 | 3.57 ± 2.74 |
| Turbidity (NTU) | Turb | 14.10 ± 7.60 | 16.24 ± 7.31 | 15.18 ± 6.36 | 15.25 ± 6.95 | 16.14 ± 7.72 | 18.23 ± 7.81 | 18.52 ± 8.65 | 15.83 ± 6.45 |
| Chlorophyll *a* (μg/L) | Chl-a | 4.20 ± 3.44 | 7.33 ± 6.68 | 4.50 ± 3.17 | 3.49 ± 2.05 | 3.11 ± 1.98 | 6.39 ± 5.68 | 7.78 ± 10.14 | 3.83 ± 2.35 |
| pH (pH unit) | pH | 5.89 ± 0.43 | 6.30 ± 0.45 | 6.42 ± 0.39 | 6.43 ± 0.29 | 6.49 ± 0.38 | 6.41 ± 0.29 | 6.48 ± 0.32 | 6.48 ± 0.34 |
| Light attenuation coefficient ($m^{-1}$) | Ke | 4.78 ± 2.52 | 4.87 ± 2.48 | 2.68 ± 1.17 | 2.58 ± 1.30 | 2.67 ± 1.27 | 2.84 ± 1.26 | 2.37 ± 0.87 | 4.35 ± 1.97 |
| Wind Speed (m/s) | WS | 0.744 ± 0.182 | 0.744 ± 0.182 | 0.744 ± 0.182 | 0.744 ± 0.182 | 0.744 ± 0.182 | 0.744 ± 0.182 | 0.744 ± 0.182 | 0.744 ± 0.182 |
| Rainfall (mm) | R | 4.318 ± 7.048 | 4.318 ± 7.048 | 4.318 ± 7.048 | 4.318 ± 7.048 | 4.318 ± 7.048 | 4.318 ± 7.048 | 4.318 ± 7.048 | 4.318 ± 7.048 |

## 3.1. Spatial Similarity with CA

Cluster analysis was applied to find out the similarity groups between the sampling stations. It produced a dendogram (Figure 2), grouping all eight sampling stations into three statistically meaningful clusters.

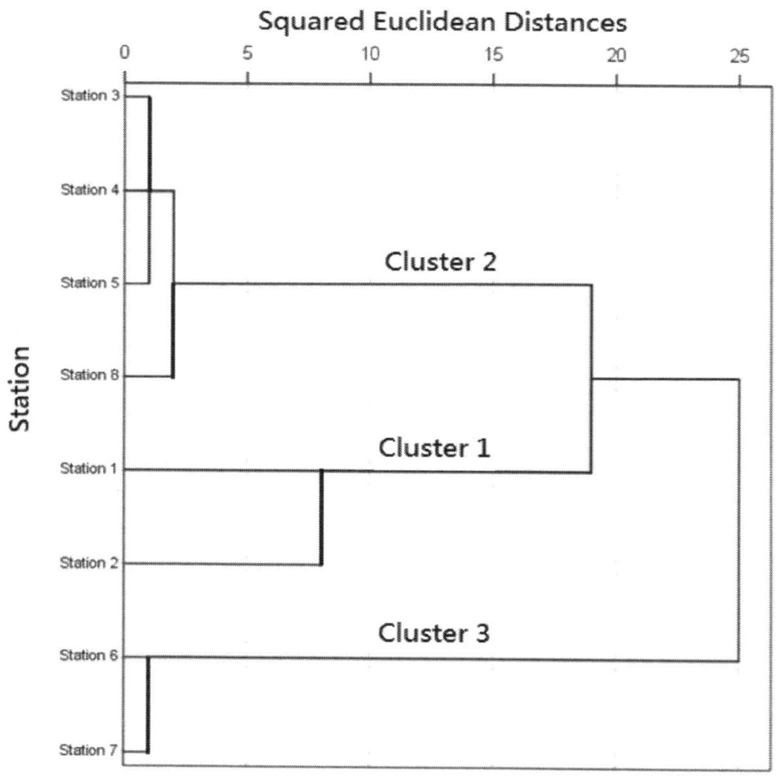


**Figure 2.** Dendrogram of cluster analysis for sampling stations accroding to water quality paramters of YYL.

The two measurement stations (1 and 2) are regarded as the cluster 2 which comprises the shallow area. Stations 3, 4, 5, and 8 are cluster 1 which corresponds to the middle water depth. Stations 6 and 7 belonging to the deep zone which constitutes cluster 3. The results show that the CA technique is useful for classification of lake waters, hence, the number of sampling sites and respective cost can be diminished in future monitoring plans. There are other reports [28–30], with similar water quality program results.

## 3.2. Principal Component Analysis and Pollution Identification

Pattern recognition of correlations among 14 parameters was best summarized by PCA/FA. The Bartlett test was used on the data set to examine the suitability of these data for PCA/FA. In this study, the covariance matrix coincided with the correlation matrix which was presented in Table 2, because FA/PCA was applied to normalized data. Overall, the correlations between variables were relatively weak. There are some positive correlations between some variables such as TP, $NH_4$-N, TN, TSS, Chl-a, and so on. The negative correlations were revealed between some variables such as DO, Temp, $NH_4$-N, TN, Chl-a, Turb, and so on. Correlation coefficients of two elements were very useful, because they numerically represented the similarity between two elements of the two water quality variables. This also indicated that PCA could successfully reduce the dimensionality of the original data set. Therefore factor analysis of the present data set further reduced the contribution of less significant variables obtained from PCA.

The Scree plot (shown in Figure 3) was applied to identify the number of PCs to be retained to understand the underlying data structure. Based on the Scree plot and the eigenvalues >1 criterion, four factors were chosen as principal factors, explaining 65.52% of the total variance in the data set. The corresponding VFs, variables loadings, eigenvalues, and explained variance are presented in Table 3.

**Table 2.** Correlation matrix of water quality parameters of YYL.

| | Temp | DO | WS | R | SD | TP | $NH_4$-N | $NO_3$-N | TN | TSS | Chl-a | Turb | pH | Ke |
|---|---|---|---|---|---|---|---|---|---|---|---|---|---|---|
| **Temp** | 1 | | | | | | | | | | | | | |
| **DO** | −0.38 ** | 1 | | | | | | | | | | | | |
| **WS** | −0.07 | −0.04 | 1 | | | | | | | | | | | |
| **R** | 0.1 | 0.1 | −0.77 ** | 1 | | | | | | | | | | |
| **SD** | −0.12 | 0.26 * | −0.02 | 0.02 | 1 | | | | | | | | | |
| **TP** | 0.24 * | −0.15 | −0.32 ** | −0.26 * | −0.27 * | 1 | | | | | | | | |
| **$NH_4$-N** | 0.30 ** | −0.27 * | −0.25 * | −0.32 ** | −0.18 | 0.37 ** | 1 | | | | | | | |
| **$NO_3$-N** | −0.26 * | −0.10 | 0.21 | 0.22 * | 0.13 | −0.28 ** | 0.04 | 1 | | | | | | |
| **TN** | 0.26 * | −0.46 ** | 0.17 | −0.21 | −0.23 * | 0.24 * | 0.35 ** | 0.16 | 1 | | | | | |
| **TSS** | 0.15 | −0.23 * | −0.33 ** | −0.32 ** | −0.18 | 0.51 ** | 0.37 ** | 0.12 | 0.25 * | 1 | | | | |
| **Chl-a** | 0.17 | −0.34 ** | −0.34 ** | 0.28 * | −0.14 | 0.39 ** | 0.48 ** | −0.10 | 0.27 * | 0.59 ** | 1 | | | |
| **Turb** | 0.36 ** | −0.48 ** | −0.18 | −0.20 | 0.01 | −0.14 | 0.55 ** | 0.16 | 0.36 ** | 0.29 ** | 0.42 ** | 1 | | |
| **pH** | 0.02 | 0.36 ** | −0.19 | 0.17 | 0.38 ** | −0.05 | −0.11 | −0.30 ** | −0.37 ** | −0.18 | −0.09 | −0.18 | 1 | |
| **Ke** | −0.11 | −0.15 | −0.01 | 0.14 | −0.38 ** | 0.04 | −0.08 | 0.07 | 0.24 * | 0.11 | 0.13 | −0.08 | −0.36 ** | 1 |

* Values are statistically significant at $p < 0.01$; ** values are statistically significant at $p < 0.05$.

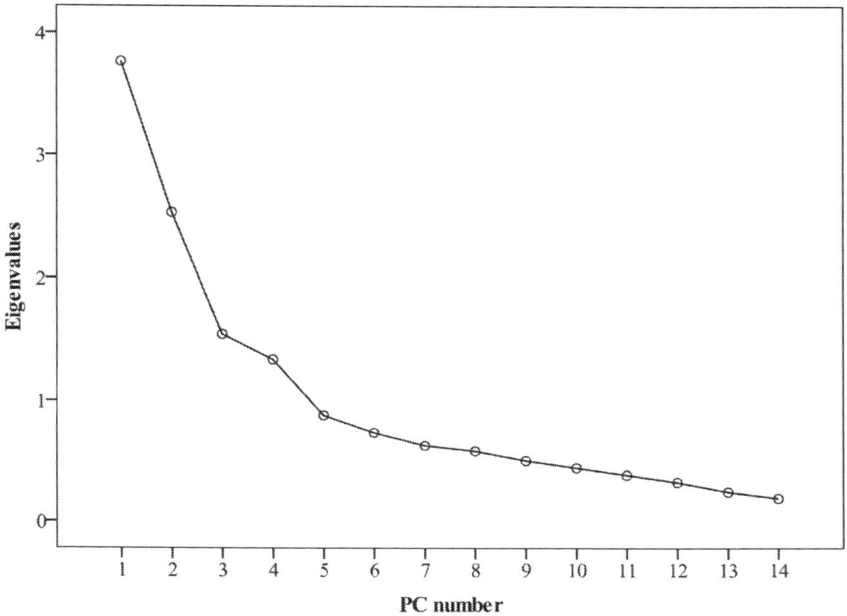


**Figure 3.** Scree plot of the characteristic roots (eigenvalues) of principal component analysis.

Liu *et al.* [31] classified the factor loadings as "strong", "moderate", and "weak", corresponding to absolute loading values of >0.75, 0.75–0.50, and 0.50–0.30, respectively. The first factor (VF1), explaining 26.89% of total variance, had moderate positive loadings on TP, $NH_4$-N, TSS, Chl-a, and Turb (turbidity). Because the $NH_4$-N concentration is a nutrient source for chlorophyll a growth, VF1 represented nitrogen source. VF2, which explained 18.08% of total variance, had a moderate positive loading on R (rainfall), WS (wind speed), TN, and pH and represents meteorological factors. VF3, explaining 11.02% of total variance, has a moderate positive loading on Ke, SD, and Turb (turbidity). This factor represents the contribution of turbidity effects in the water column. VF4, explaining 9.54% of total variance, had a moderate positive loading on $NO_3$-N and water temperature and represented the nitrate factor. The analyzed results revealed that FA/PCA can serve as an important means to identify the main factors affecting water quality in the alpine lake.

**Table 3.** Loading of 14 parameters on significant VFs for water quality data set.

| Parameters | Four significant PCs | | | |
|---|---|---|---|---|
| | VF1 | VF2 | VF3 | VF4 |
| Temp | 0.465 | 0.038 | 0.309 | −0.623 |
| DO | −0.582 | 0.437 | −0.205 | 0.171 |
| WS | −0.409 | −0.696 | 0.201 | −0.237 |
| R | 0.383 | 0.735 | −0.105 | 0.218 |
| SD | −0.367 | 0.330 | 0.581 | 0.254 |
| TP | 0.610 | 0.224 | −0.309 | −0.218 |
| $NH_4$-N | 0.718 | 0.096 | 0.252 | 0.051 |
| $NO_3$-N | −0.043 | −0.460 | 0.299 | 0.704 |
| TN | 0.536 | −0.543 | 0.118 | −0.105 |
| TSS | 0.698 | 0.111 | −0.163 | 0.310 |
| Chl-a | 0.737 | 0.133 | −0.047 | 0.148 |
| Turb | 0.655 | −0.067 | 0.533 | 0.110 |
| pH | −0.314 | 0.649 | 0.217 | −0.214 |
| Ke | 0.162 | −0.429 | −0.627 | 0.132 |
| Eigenvalue | 3.76 | 2.53 | 1.54 | 1.34 |
| Percentage of total variance | 26.89 | 18.08 | 11.02 | 9.54 |
| Cumulative percentage of variance | 26.89 | 44.96 | 55.98 | 65.52 |

## 3.3. Geostatistical Mapping

Geostatisitcal techniques were used for the mapping of principle components and factor scores over the study area. Due to the long period between each observation campaign, the temporal correlation among the observations is assumed to be ignorable in this analysis. Table 4 shows that the spatial dependence structure varies across the identified contributing factors by the common multivariate analysis. It implies the variation of spatial patterns of impacts to water quality from the contributing factors. Among them, the impact of nitrogen nutrients changes more significantly over space than other contributing factors. The experimental and modeled variograms of PC1 and FA1 are shown in Figure 4. The variogram figure in time for PC1 and FA1 is also presented in Figure 5. It is clear that the variogram value approximates to sill in cases of the temporal lags in month among observations larger than 0. It implies the low correlation between the observations collected in different months. The contaminants from nitrogen nutrient are more localized as shown in Figure 6. On the other hand, the effects from the sunlight, organic matter, and nitrate nutrition present much smoother variations across the study area. This implies the sources of these contributors are more homogeneously distributed over the lake. It is noticeable that the range of the semivariogram model of second principle component is excessively larger than those of the models of other factors. It implies that the meteorological effects derived from PCA contribute a relatively large scale variation of water quality in space with respect to the scale of the study area.

The spatial distribution of the PC and FA can vary over time. Our analysis shows that the spatial distributions of PC (or FA) of the observations collected in the same month are generally similar. As for the PC obtained at different months, their spatial distribution can be distinct. This variability can result from meteorological condition and physico-chemical characteristics.

The general characteristics can be seen in Figure 7 in which a clear increasing trend from south to north of principle component is shown.

## 4. CONCLUSIONS

Water quality data collected from eight monitoring stations located around the subtropical alpine Yuan-Yang Lake in Taiwan have been examined by unsupervised pattern recognition (CA) and display methods (PCA/FA) to yield correlations between variables and water quality similarity in the lake. Cluster analysis confirmed the existence of three types of water quality (*i.e.*, shallow, middle and deep zones of the lake). The PCA and FA assisted to extract and recognize the factors or origins responsible for water quality variations. PCA/FA identified four latent factors that explained 65.52% of total variance, namely nitrogen source, meteorological factor, turbidity effect, and nitrate factor, respectively. Geostatisitcal techniques were used for the mapping of principle components and factor scores in the lake. The results revealed that the impact of nitrogen nutrients changes more significantly over space than other contributing factors. It means that nitrogen sources consist of important contribution to affect the water quality of the lake.

Thus, this study illustrated the usefulness of multivariate statistical and geostatistical techniques for the analysis and interpretation of complex data set, water quality assessment, and identification of important contribution in nutrient source in the YYL.

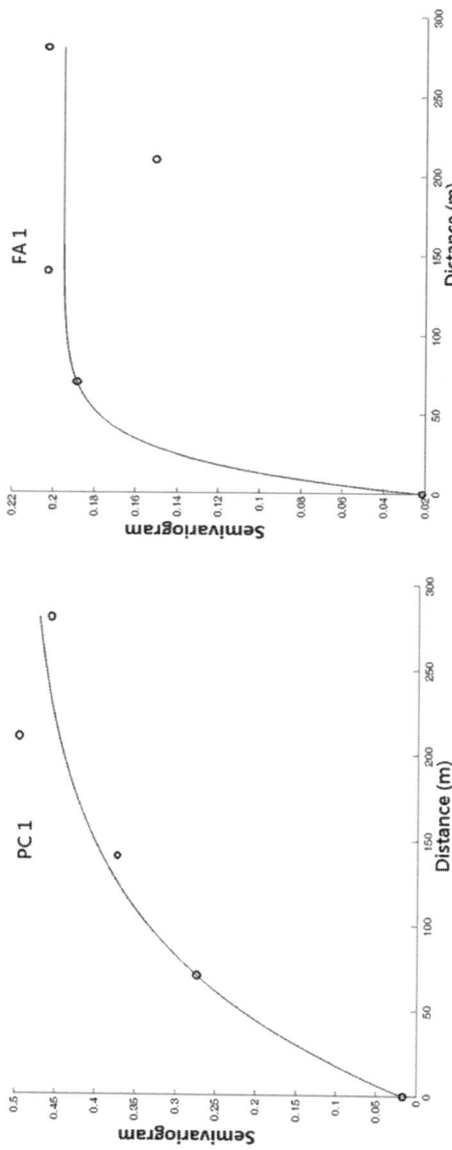


**Figure 4.** The experimental and modeled variograms of PC1 and FA1.

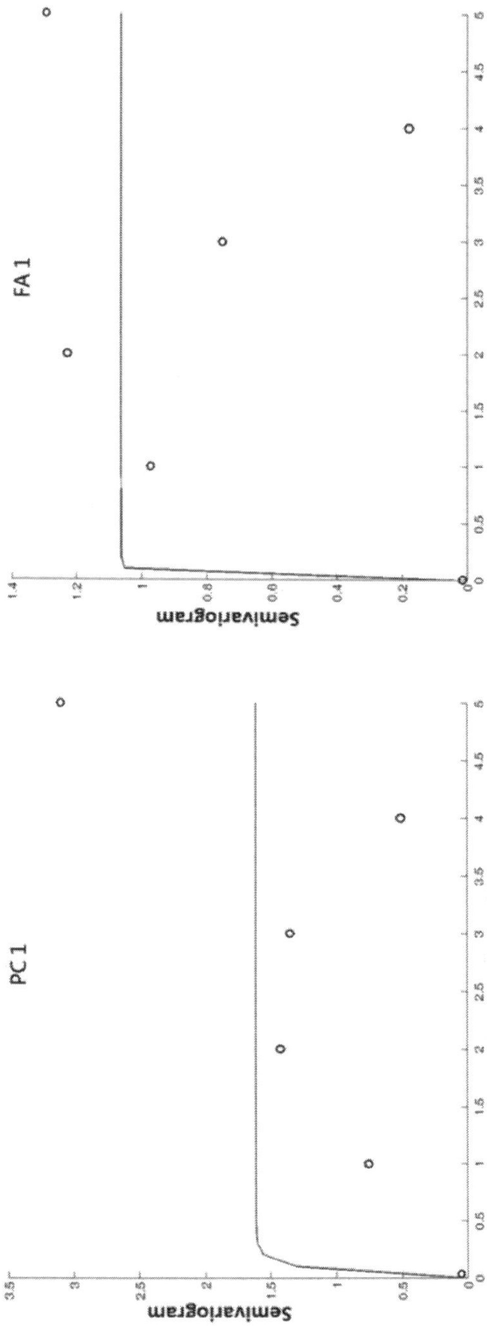


**Figure 5.** Variograms in time for PC1 and FA1.

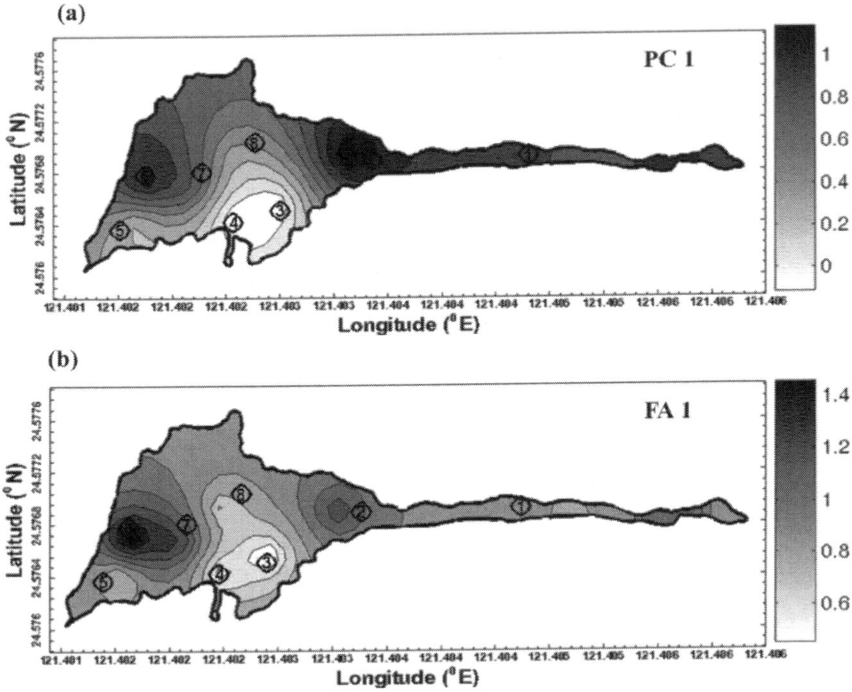


**Figure 6.** Spatial distribution of **(a)** first principle component and **(b)** first factor score at the time on the measured data of September 12, 2009.

**Table 4.** Variogram models used for spatial mapping.

| Variables | Variogram models |
|---|---|
| PC1 | Nugget[0.031] + Exponential[0.466, 287.106] |
| PC2 | Nugget[0.007] + Gaussian[5.004, 1116.3] |
| PC3 | Nugget[0.036] + Gaussian[1.443, 259.137] |
| PC4 | Nugget[0.018] + Gaussian[0.305, 215.136] |
| FA1 | Nugget[0.038] + Exponential[0.157, 65.983] |
| FA2 | Nugget[0.003] + Exponential[0.080, 185.810] |
| FA3 | Nugget[0.010] + Gaussian[2.056, 383.165] |
| FA4 | Nugget[0.010] + Gaussian[0.409, 288.627] |

Note: The notations that Nugget[ $s_1$ ] + Exponential(or Gaussian)[ $s_2$ , $r_2$ ] denote the nest model of nugget model effect of sill $s_1$ and exponential (or Gaussian) model of sill $s_2$ and range $r_2$ in meters. PC: Principal Component; FA: Factor Analysis.

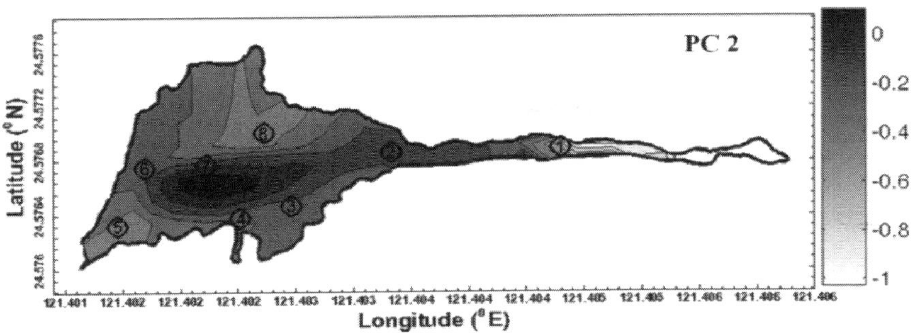


**Figure 7.** Spatial distribution of second principle component by ordinary kriging method on the measured data of February 14, 2009.


## ACKNOWLEDGMENTS

This study was supported by the National Science Council and Academia Sinica, Taiwan, under the grant number NSC-96-2628-E-239-012-MY3 and AS-98-TP-B06, respectively. The financial support is highly appreciated.


## REFERENCES


1. Zeng, X; Rasmussen, TC. Multivariate statistical characterization of water quality in Lake Lanier, Georgia, USA. *J. Environ. Qual* **2005**, *34*, 1980–1991.

2. Heikka, RA. Multivariate monitoring of water quality: A case study of Lake Simple, Finland. *J. Chemomet* **2007**, *22*, 747–751.

3. Nakasone, H. Effect on water quality in irrigation reservoir due to application reduction of nitrogen fertilizer. *Paddy Water Environ* **2009**, *7*, 65–70.

4. Palma, P; Alvarenga, P; Palma, VL; Fernandes, RM; Soares, AMVM; Barbosa, IR. Assessment of anthropogenic sources of water pollution using multivariate statistical techniques: A case study of Alqueva's reservoir, Portugal. *Environ. Monit. Assess* **2010**, *165*, 539–552.

5. Singh, KP; Malik, A; Mohan, D; Sinha, S. Multivariate statistical techniques for the evaluation of spatial and temporal variations in water quality of Gomti river (India): A case study. *Water Res* **2004**, *38*, 3980–3992.

6. Li, R; Dong, M; Zhao, Y; Zhang, L; Cui, Q; He, W. Assessment of water quality and identification of pollution sources of plateau lakes in Yunnan (China). *J. Environ. Qual* **2007**, *36*, 291–297.

7. Kazi, TG; Arain, MB; Jamali, MK; Jalbani, N; Afridi, HI; Sarfraz, RA; Baig, JA; Shah, AQ. Assessment of water quality of polluted lake using multivariate statistical techniques: A case study. *Ecotox. Environ. Safe* **2009**, *72*, 301–309.

8. Singh, KP; Malik, A; Sinha, S. Water quality assessment and apportionment of pollution sources of Gomti river (India) using multivariate statistical techniques: A case study. *Anal. Chim. Acta* **2005**, *538*, 355–374.

9. Kim, JH; Choi, CM; Kim, SB; Kwun, SK. Water quality monitoring and multivariate statistical analysis for rural streams in South Korea. *Paddy Water Environ* **2009**, *7*, 197–208.

10. Chehata, M; Jasinski, D; Monteith, MC; Samuels, WB. Mapping three-dimensional water-quality data in the Chesapeake Bay using geostatistics. *J. Am. Water Resour. Assoc* **2007**, *43*, 813–828.

11. Todd, MJ; Lowrance, RR; Goovaerts, P; Vellidis, G; Pringle, CM. Geostatistical modeling of the spatial distribution of sediment oxygen demand within a Coastal Plain backwater watershed. *Geoderma* **2010**, *159*, 53–62.

12. Lopez-Granados, F; Jurado-Exposito, M; Pena-Barragan, JM; Garcia-Torres, L. Using geostatistical and remote sensing approaches for mapping soil properties. *Eur. J. Agron* **2005**, *23*, 279–289.

13. Nour, MH; Smit, DW; El-Din, MG. Geostatistical mapping of precipitation: Implication for rain gauge network design.*Water Sci. Technol* **2006**, *53*, 101–110.

14. Sauquet, E. Mapping mean annual river discharge: Geostatistical development for incorporating river network dependencies. *J. Hydrol* **2006**, *331*, 300–314.

15. Wackernagel, H; Lajaunie, C; Blond, N; Vautard, R. Geostatistical risk mapping with chemical transport model output and ozone station data. *Ecol. Model* **2004**, *179*, 177–185.

16. USEPA. *Methods for Chemical Analysis of Water and Waste*; US Environmental Protection Agency: Washington, DC, USA, 1983.

17. Wunderlin, DA; Diaz, MP; Ame, MV; Pesce, SF; Hued, AC; Bistoni, MA. Pattern recognition techniques for the evaluation of spatial and temporal variations on water quality. A case study: Suquira river basin (Cordoba-Argentina).*Water Res* **2001**, *35*, 2881–2894.

18. Simeonov, V; Stratis, JA; Samara, C; Zachariadis, G; Vousta, D; Anthemidis, A; Sofoniou, M; Kouimtzis, Th. Assessment of the surface water quality in Northern Greece. *Water Res* **2003**, *37*, 4119–4124.

19. Kowalkowski, T; Zbytniewski, R; Szpejna, J; Buszewski, B. Application chemometrics in river water classification.*Water Res* **2006**, *40*, 744–752.

20. Helena, B; Pardo, R; Vega, M; Barrado, E; Fernandez, JM; Fernandez, L. Temporal evolution of groundwater composition in an alluvial aquifer (Pisuerga river, Spain) by principal component analysis. *Water Res* **2000**, *34*, 807–816.

21. Brumelis, G; Lapina, L; Nikodemus, O; Tabors, G. Use of an artificial model of monitoring data to aid interpretation of principal component analysis. *Environ. Modell. Softw* **2000**, *15*, 755–763.

22. Abdul-Wahab, SA; Bakheit, CS; Al-Alawi, SM. Principal component and multiple regression analysis in modelling of ground-level ozone and factors affecting its concentration. *Environ. Modell. Softw* **2005**, *20*, 1263–1271.

23. *STATISTICA (data analysis software system), Version 8*; Statsoft INC: Tulsa, OK, USA, 2007.

24. Cressie, N. Fitting variogram models by weighted least-squares. *J. Int. Assoc. Math. Geol* **1985**, *17*, 563–586.

25. Goovaerts, P. *Geostatistics for Natural Resources Evaluation*; Oxford University Press: New York, NY, USA, 1977.

26. Wackernagel, H. *Multivariate Geostatistics: An Introduction with Applications*; Springer: Berlin, Germany, 2003.

27. Yu, HL; Kolovos, A; Christakos, G; Chen, JC; Warmerdam, S; Dev, B. Interactive spatiotemporal modelling of health systems: The SEKS–GUI framework. *Stoch. Environ. Res. Risk Assess* **2007**, *21*, 555–572.

28. Kim, JH; Kim, RH; Lee, J; Cheong, TJ; Yum, BW; Chang, HW. Multivariate statistical analysis to identify major factors governing groundwater quality in the coastal area of Kimje, South Korea. *Hydrol. Process* **2005**, *19*, 1261–1276.

29. Shrestha, S; Kazama, F. Assessment of surface water quality using multivariate statistical techniques: A case study of the Fuji river basin, Japan. *Environ. Modell. Softw* **2007**, *22*, 464–475.

30. Zaharescu, DG; Hooda, PS; Soler, AP; Fernandez, J; Burghelea, CI. Trace metals and their source in the catchment of the high

altitude Lake Respomuso, Central Pyrenees. *Sci. Total Environ* **2009**, *407*, 3546–3553.

31. Liu, CW; Lin, KH; Kuo, YM. Application of factor analysis in the assessment of groundwater quality in a Blackfoot disease area in Taiwan. *Sci. Total Environ* **2003**, *313*, 77–89.

6

# INVESTIGATION OF SPATIAL AND TEMPORAL TRENDS IN WATER QUALITY IN DAYA BAY, SOUTH CHINA SEA

**Mei-Lin Wu [1,2], You-Shao Wang [1,2,*], Jun-De Dong [1], Cui-Ci Sun [1,2], Yu-Tu Wang [1,2], Fu-Lin Sun [1,2] and Hao Cheng [1,2]**

[1]State Key Laboratory of Oceanography in the Tropics, South China Sea Institute of Oceanology, Chinese Academy of Sciences, Guangzhou 510301, China

[2]Marine Biology Research Station at Daya Bay, Chinese Academy of Sciences, Shenzhen 518121, China

## ABSTRACT

The objective is to identify the spatial and temporal variability of the hydrochemical quality of the water column in a subtropical coastal system, Daya Bay, China. Water samples were collected in four seasons at 12 monitoring sites. The Southeast Asian monsoons, northeasterly from October to the next April and southwesterly from May to September have also an important influence on water quality in Daya Bay. In the spatial pattern, two groups have been identified, with the help of multidimensional scaling analysis and cluster analysis. Cluster I consisted of the sites S3, S8, S10 and S11 in the west and north coastal parts of Daya Bay. Cluster I is mainly related to anthropogenic activities such as fish-farming. Cluster II consisted of the rest of the stations in the center, east and south parts of Daya Bay. Cluster II is mainly related to seawater exchange from South China Sea.

## KEYWORDS

Daya Bay; principal component analysis; multidimensional scaling analysis; statistical techniques; water quality

## 1. INTRODUCTION

Daya Bay lies on the southern coast of China (from 22°31′12″ to 22°50′00″ N, 114°29′42″ to 114°49′42″ E) like a door shape, with a coastal line of 92 km, and an area of 600 $km^2$. The water depth ranges from 6 to 16 m, with an average water depth of 10 m [1]. There are two towns (Dapeng Town and Nanao Town) in the western coast area. The north and east coast area belong to Huizhou, China, there are five towns, namely Xiachong Town, Aotou Town, Renshan Town, Xunliao Town and Tiechong Town. In recent years, the rapid economic development and anthropogenic activities from Shenzhen and Huizhou have had a great influence on the environment of this bay. It has been one of the main aquaculture areas in Guangdong Province, due to its excellent water quality. Daya Bay Nuclear Power Station (DNPS) and Lingao Nuclear Power Station (LNPS) have also been operated there. The stronger northeast monsoon prevails in the area from October to April; and the Southeast Asian monsoons predominate southwesterly from May to September.

Some studies have been conducted on the environmental effects caused by anthropogenic activities in the bay. Some authors have described phytoplankton biomass and primary production [2], high temperature effluent from DNPS [3], nutrients in aquaculture area [4] and nutrients [5–8]. Other studies have been focused on water quality [9–11], organic pollutants in water or surface sediments [12], and heavy metals [13,14].

In the present work, we have done a relatively comprehensive study of the water quality of Daya Bay. Besides, multivariate statistical techniques were used to establish the effects caused by the different human activities and seasonal variations found in the bay.

# 2. MATERIALS AND METHODS

## 2.1. Sampling and Analytical Methods

Seawater samples for analysis of nutrients, chlorophyll *a* were taken using 5-L GO FLO bottles at surface and bottom layers of all stations (Figure 1) in January (Winter), April (Spring), August (Summer) and November (Autumn) in 2003 (GB12763-91, China). The determinations of water temperature, and salinity were performed *in situ* by the water quality Monitoring System. Water samples from various depths were analyzed for nitrate and nitrite with a SKALAR auto-analyzer (Skalar Analytical B.V. SanPlus, The Netherlands). Ammonium and total phosphorus (TP) were analyzed by the methods of oxidation with hypobromite and molybdophosphoric blue using an UV1601 spectrophotometer (Shimadzu Corporation, Japan), respectively. Chemical oxygen demand (COD), 5-day bio-chemical oxygen demand ($BOD_5$) and transparency (Secchi) were tested according to "The specialties for marine monitoring" (GB17378.4-1998, China) and Wang *et al.* [8]. Dissolved oxygen (DO) was determined with the Winkler titration method. DIN is the bsum of nitrate and nitrite and ammonium.

Two replicates of 1.5 L samples from the depths mentioned above were filtered through 47 mm GF/F filters and were deep frozen immediately at −20 °C. At the end of the cruise, all filters were transported to the shore laboratory in liquid nitrogen. Within a week, the chlorophyll *a* (Chl *a*) was extracted in10 mL 90% acetone in the dark for 24 h in a refrigerator and its concentration was determined with 10-AU Fluorometry (Turner Designs, USA).

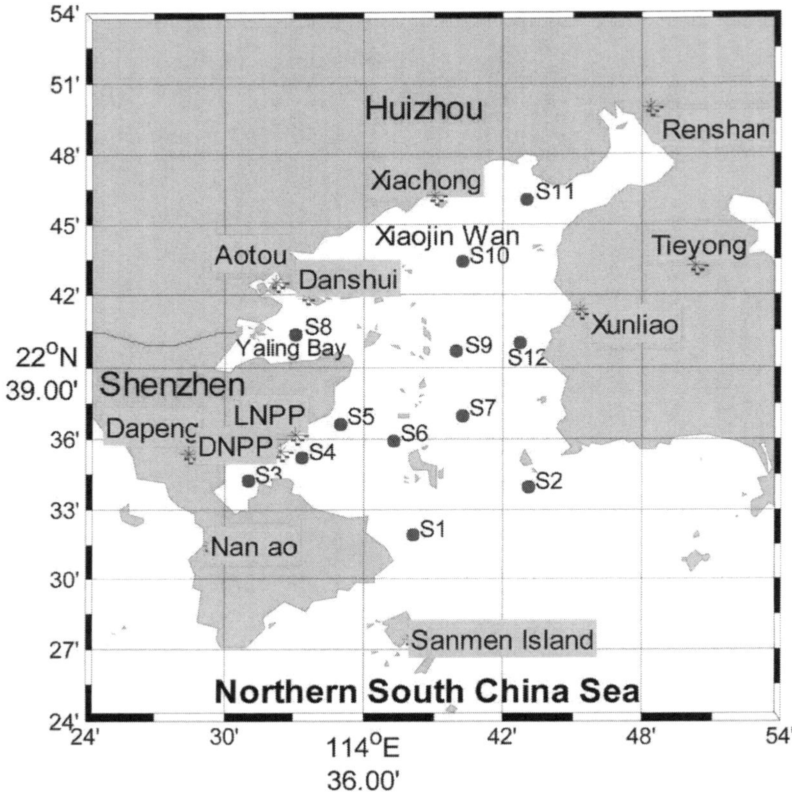


**Figure 1.** Monitoring stations in Daya Bay [10].

## 2.2. Data Treatment

The data for this study are the mean data obtained from surface and bottom waters. In this study, well known multivariate statistical techniques such as principal component analysis (PCA) and multidimensional scaling (MDS) were used to identify the anthropogenic and natural effects on the water environment status. PCA was used to understand the correlation structures between variables and to combine them into groups to reduce the dimensions of the variables for making decisions and interpretations easier. MDS analysis was used to explore the similarities or dissimilarities (distances) in water quality between the surface water quality monitoring stations in the bay.

Data was auto-scaled in order to avoid misclassification due to wide differences in data dimensionality. The data were normalized with mean and variance of zero and one, respectively. All the procedures were performed using MATLAB R 2008b (Mathworks Inc., USA).

# 3. RESULTS

Two-way analysis of variance (ANOVA) was performed in order to show the existence of temporal and spatial differences between water samples from Daya Bay. Results from ANOVA did not show any temporal variations of COD obtained from different sampling campaigns but showed significant temporal change of other variables, such as temperature, salinity, Secchi, DO, $BOD_5$, Chl *a*, DIN and TP. Significant spatial differences were found for all the variables including temperature, salinity, Secchi, DO, $BOD_5$, COD, Chl *a*, DIN and TP. This result may be explained by human activities and natural influences.

The horizontal distributions of the surface temperature are similar in all four seasons, with higher values in the western and north parts of the bay than in other parts of the bay (Figure 2). The horizontal distributions of the surface salinity are similar in four seasons, the surface salinity decrease from the mouth to the top in the bay in summer; it is higher in the eastern and northern part than that in western part in winter (Figure 3). In summer, the salinity is lower than that in other seasons; the rain is an important factor, which diminishes the surface salinity.

## 3.1. Principal Component Analysis

The particular challenge in the case of water quality monitoring is the complexity associated with analyzing the large number of measured variables. Therefore, in this study, principal component analysis, which is a well known data reduction technique, was utilized to extract the main components of the water quality data.

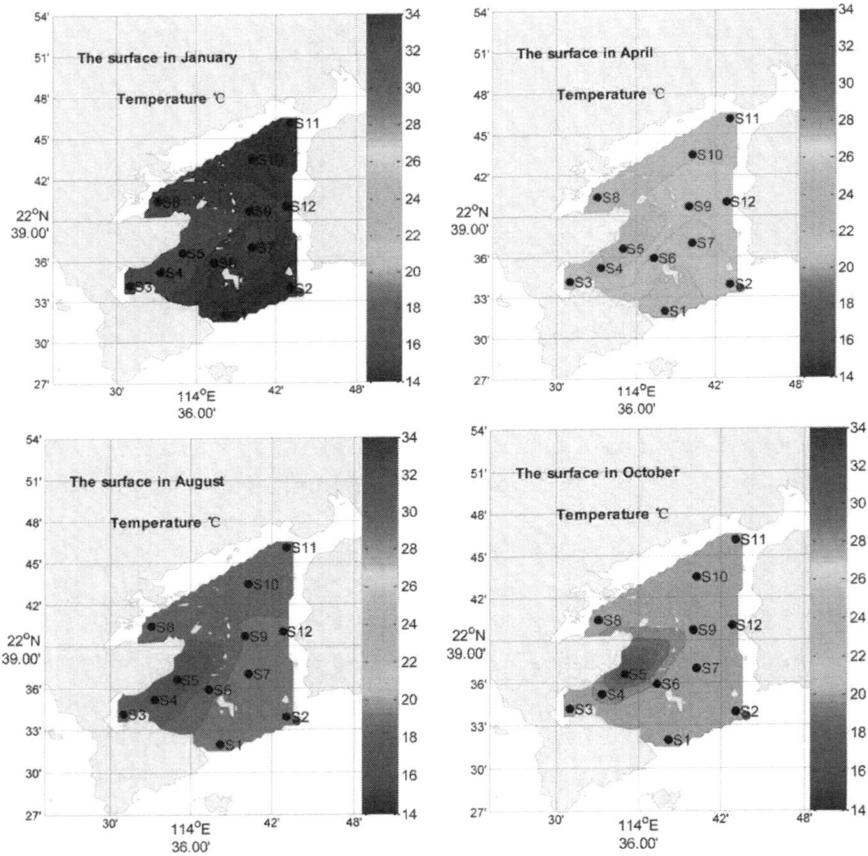


**Figure 2.** The spatial distribution of surface temperature in four seasons, respectively.

The loadings of the three retained principal components (PCs) with classical PCA are shown in Table 1. PC1 (35.1039% of the variance) is mainly related to Secchi, $BOD_5$, COD and DIN. PC2 (15.3675% of the variance) is characterized by Chl *a* and TP. PC3 explains 15.3675% of the variance, related mainly to DO.

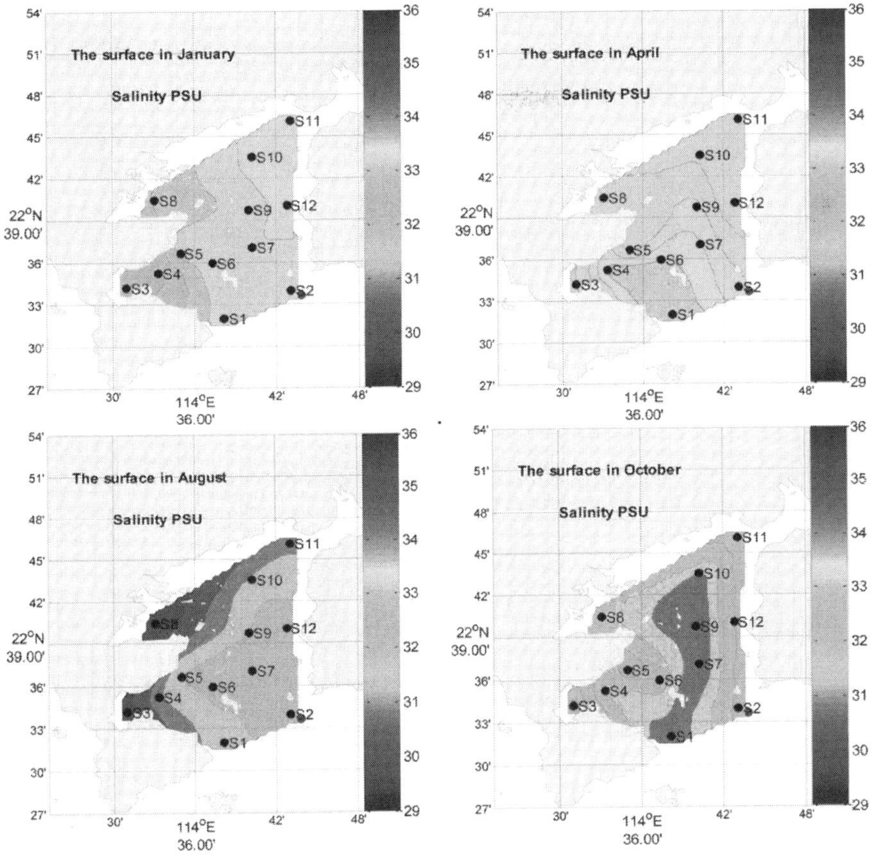


**Figure 3.** The spatial distribution of surface salinity in four seasons, respectively.

The matrix of scores provides information about the distribution of patterns or sources of contamination among samples. The matrix of loadings defines the contribution of the original variables to each one of these contamination patterns or sources. The data are distributed in a limited region of space spanned by the two well-defined PC axes. Seasonal pattern has been demonstrated by the scores of the twelve stations in four seasons. The seasonal stations are separated in the plane; the seasonal character has distinct. The border of the four groups is clearly marked in Figure 4. The principal components score plot can not only interpret the spatial distribution by clustering the samples, but can also describe their different characteristics and help to find out the relationship between different parameters (variables) by the parameter lines.

**Table 1.** Loadings of 7 physicalechemical parameters on the seven PCs.

| | PC1 | PC2 | PC3 | PC4 | PC5 | PC6 | PC7 |
|---|---|---|---|---|---|---|---|
| **Secchi** | −0.4636 | −0.1119 | −0.2607 | 0.2654 | −0.7799 | 0.1246 | 0.1021 |
| **DO** | −0.2384 | 0.3932 | −0.6293 | 0.1145 | 0.2366 | −0.5657 | −0.0585 |
| **$BOD_5$** | 0.5016 | −0.3028 | −0.3602 | 0.2290 | 0.0282 | −0.0341 | 0.6874 |
| **COD** | 0.4370 | −0.3401 | −0.4570 | −0.2205 | −0.2120 | 0.0193 | −0.6250 |
| **Chl *a*** | 0.2433 | 0.5423 | −0.1815 | 0.4548 | 0.0705 | 0.6141 | −0.1577 |
| **DIN** | 0.4332 | 0.2497 | 0.3955 | 0.3629 | −0.4038 | −0.5310 | −0.1295 |
| **TP** | 0.1971 | 0.5187 | −0.0898 | −0.6905 | −0.3493 | 0.0620 | 0.2851 |
| **Eigenvalue** | 2.4573 | 1.9895 | 1.0757 | 0.6068 | 0.5364 | 0.2769 | 0.0572 |
| **Variance (%)** | 35.1039 | 28.4218 | 15.3675 | 8.6692 | 7.6634 | 3.9563 | 0.8179 |
| **Cumulative (%)** | 35.1039 | 63.5257 | 78.8932 | 87.5624 | 95.2258 | 99.1821 | 100.0000 |

The parameter lines were obtained from the factor loadings of the original variables. They stand for the contribution (significance) of the parameters to the samples. The closer two-parameter lines lie together, the stronger the mutual correlation is. For example, TP and Chl *a* have the highest positive correlation coefficient (0.95), so the two lines are very close. A very weak or no correlation between the two parameters was found, angle between two lines was about 90°. Similarly, an obtuse angle between two lines represents a negative correlation, *etc.*, DO and Chl *a*.

From the principal components score plot, the temporal variations of the samples in the bay can be clearly observed. In the first quadrant, a cluster of the samples in winter is characterized by high TP, DIN and Chl *a*, as these parameter lines are located in this quadrant. As mentioned before, external pollution sources from land-resources could be introduced into the bay by river, runoff and rain, *etc.*

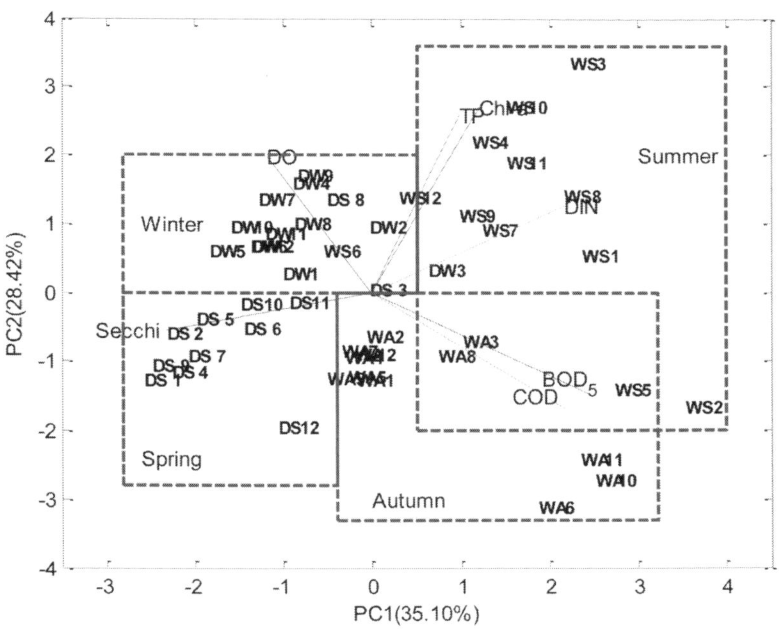


**Figure 4.** The loadings of variables and scores of the 12 stations for the first two PCs (DW-Winter, DS-Spring, WS-Summer and WA-Autumn), respectively. The number denotes the station number; the letter denotes the variable.

## 3.2. Multidimensional Scaling Analysis

In this section, the water quality characteristics data were analyzed by using two-dimensional and three-dimensional MDS analysis to identify similarities and differences between surface water quality monitoring stations. STRESS (Standardized Residual Sum of Squares), which is used to evaluate how well a particular configuration reproduces the observed distance matrix, was found to be 0.00008 even for two-dimensional MDS. Because, STRESS values close to zero shows that the "fit" is almost perfect and the results of the MDS analysis is reasonable and reliable.

Temporal MDS was performed on raw data after dividing the whole data set into four (winter, spring, summer and autumn) seasonal groups (Figures 5 to 8). The 12 surface water quality monitoring stations according to Dimension-1 and Dimension-2 via MDS analysis had different spatial patterns in four seasons, respectively (Figures 5 to 8). In Winter, where all the 12 sampling sites on the bay were grouped in to three clusters with cluster 1 with grouping four sites (S1, S2, S4–S7, S9, S10 and S12), cluster 2 with another two sites (S8 and S11), and cluster 3 comprising site (S3). The circulation has different characteristics in different seasons, and land-resources inputs have been different. Thus, MDS was performed on the mean values of variables. The 12 surface water quality monitoring stations were grouped into two clusters (Figure 9). Cluster I consisted of S3, S8, S10 and S11 in the west and north coastal parts of Daya Bay, and cluster II consisted of the rest stations in the center, east and south parts of Daya Bay.

# 4. DISCUSSION

Daya Bay is a semi-closed bay and has several seasonal creeks. In the south, Daya Bay exchanges its water with the South China Sea. Therefore, the water in South China Sea plays an important role on that in Daya Bay. Furthermore, Daya Bay is located in a subtropical zone, where the stronger northeast monsoon prevails from October to April; and the southwesterly Southeast Asian monsoons predominate from May to September. The seasonal change character has been proved the dry and wet seasonal characters in Daya Bay [7]. The average precipitation is 1,827 mm in Daya Bay, the maximum rainfall is 370 mm in June, and the minimum rainfall is 30 mm in December [15]. There is plenty of rain from May to October (about 80% precipitation of the whole year), and there is less from November to the next April (about 20% precipitation of the whole year). The salinity presents the most significant temporal differences ($p = 0.00$). In Summer, fresh water input from several seasonal rivers diminishes the surface salinity, with lower salinity in west north part of the bay. The salinity may be strongly related to the runoff and precipitation.

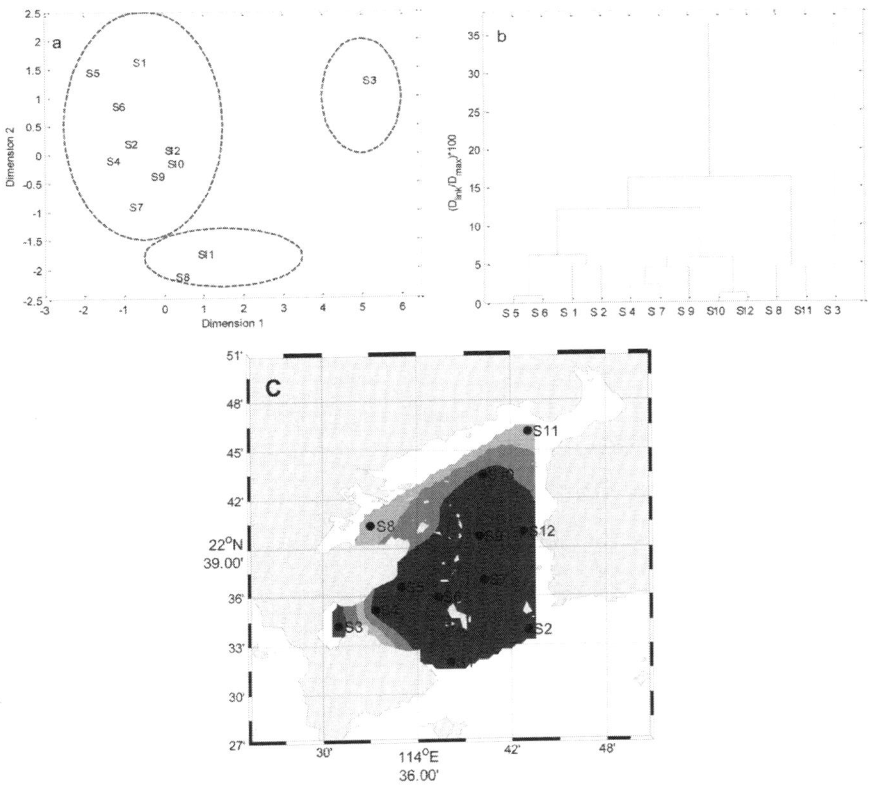


**Figure 5.** The results of multidimensional scaling analysis and cluster analysis: (**a**) Multidimensional scaling analysis plot for the monitoring stations in Winter; (**b**) Dendrogram based on Ward's method for monitoring stations in Winter; (**c**) Map of the resulting zones of Daya Bay in Winter.

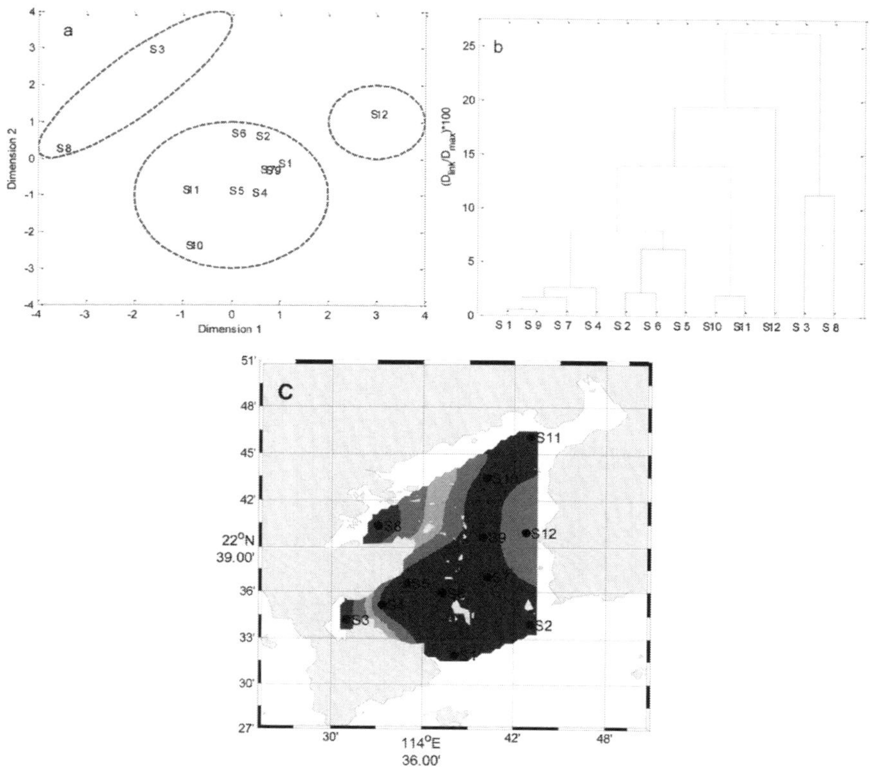


**Figure 6.** The results of multidimensional scaling analysis and cluster analysis: (**a**) Multidimensional scaling analysis plot for the monitoring stations in Spring; (**b**) Dendrogram based on Ward's method for monitoring stations in Spring; (**c**) Map of the resulting zones of Daya Bay in Spring.

Temperature has a distinct seasonal change (Figure 2).Temperature is an important climatic factor, it can be an important indicator of natural processes in the subtropical area [7]. The result of analysis of variance analysis also shows that the temperature presents the most significant temporal difference ($p$ = 0.00). The spatial distribution of the water temperature showed high values in the western part of Daya Bay near the nuclear power stations (near S5) and low values in the eastern and southern parts of the Bay. Warm discharge from DNPS could lead to the disadvantaged growth of some marine species relative to others [16].

With a 1~3 °C change of temperature, the discharge plume from DNPS may also result in some ecological changes in the lower portion of Daya Bay, particularly in Dapeng Cove [3].

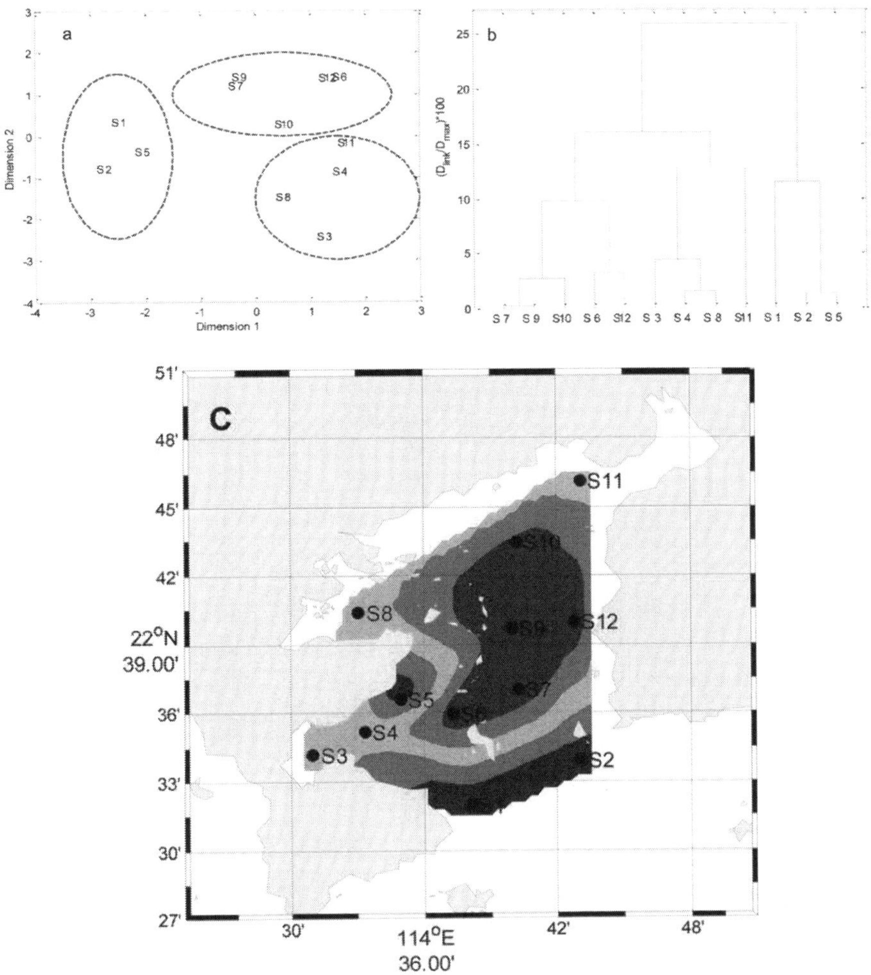


**Figure 7.** The results of multidimensional scaling analysis and cluster analysis: (**a**) Multidimensional scaling analysis plot for the monitoring stations in Summer; (**b**) Dendrogram based on Ward's method for monitoring stations in Summer; (**c**) Map of the resulting zones of Daya Bay in Summer.

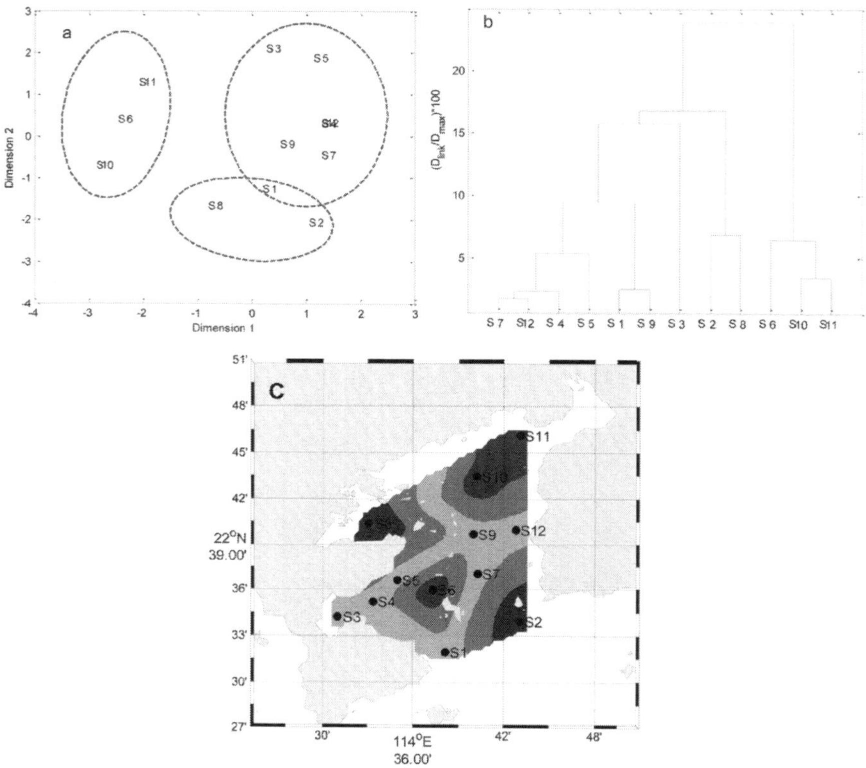

**Figure 8.** The results of multidimensional scaling analysis and cluster analysis: (**a**) Multidimensional scaling analysis plot for the monitoring stations in Autumn; (**b**) Dendrogram based on Ward's method for monitoring stations in Autumn; (**c**) Map of the resulting zones of Daya Bay in Autumn.

The stratification due to temperature and salinity differences between surface and bottom waters within the Bay started to develop in June, was strongest from July to September, and disappeared in October. The water of lower temperature and high salinity intrudes into the bay along the bottom from the South China Sea under the influence of the weak monsoons southwesterly from May to September [15,17]. The stratified waters were distinct in Summer. Temperature and salinity in the Bay were uniform with depth from November until April of the following year [18]. The waters are vertically mixed in Daya Bay under the influence of the northeast monsoon [19,20]. In addition, the circulation have different pattern all through year. The clockwise Euler Residual Current in Spring, Summer and Autumn in Daya Bay [1] carried the nutrients from land-based water from the west and north in the bay through this area, meanwhile, tidal current carried the nutrients from the South China Sea through this area as well, which may cause the higher concentration of Chl *a* in this area [5]. The highest phytoplankton density was at station S8 near Aotou harbor and the cage culture areas in the north-western part of Daya Bay. This result is consistent with the higher nutrient levels near station S3 and S8. Nutrients could play an important role in determining the density of phytoplankton in Daya Bay [18]. The mean concentrations of Chl *a* in monitoring stations (S3, S8, S10 and S11) and the rest stations (S1, S2, S4–S7, S9 and S12) are 4.70 $mg{\cdot}L^{-1}$ and 2.29 $mg{\cdot}L^{-1}$, respectively. The west and north part of the bay (Dapeng Cove and Aotou) may have higher nutrient concentration than elsewhere [21]. The nutrients play an important role on the water quality in the marine aquaculture area and anthropogenic sources of pollution (S3, S8, S10 and S11).

From the results of the analysis of water balances during both two monsoons periods by box models and mass conservation arguments, the water residence time is shorter during the southwest monsoon period than during the northeast monsoon period [1]. Because the turnover time of seawater within the system favors the operation of different internal processes (e.g., biogeochemical, nutrient assimilation, sediment resuspension) in the water column and changes its own characteristics [22].

Annually, the two clusters have been identified by MDS and CA (Figure 9). The monitoring stations (S3, S8, S10 and S11) in the west and north coastal parts of Daya Bay are together clustered. The rest stations (S1, S2, S4–S7, S9 and S12) in the center, east and south parts of Daya Bay are grouped. The S3 and S8 lie in Dapeng Cove and the northwest part near Aotou harbor, respectively. These two bays are used for fish-farming. The human being activities have a strong influence on aquatic environment in the west and north coastal parts of Daya Bay [23]. The mean concentrations of $BOD_5$ in monitoring stations (S3, S8, S10 and S11) and the rest stations (S1, S2, S4–S7, S9 and S12) are 1.73 $mg{\cdot}L^{-1}$ and 1.28 $mg{\cdot}L^{-1}$, respectively. The mean concentrations of COD in monitoring stations (S3, S8, S10 and S11) and the rest stations (S1, S2, S4–S7, S9 and S12) are 1.04 $mg{\cdot}L^{-1}$ and 0.80 $mg{\cdot}L^{-1}$, respectively. The concentrations of the parameters related to anthropogenic pollution like $BOD_5$and COD are higher in monitoring stations (S3, S8, S10 and S11) than that in the rest stations. This result indicates that contamination is mainly from coast human activities and marine aquiculture. The water exchange is also crucial for the balance between continental inputs delivered to the coastal region and coastal water inflow. The Southeast Asian monsoons, northeasterly from October to April and southwesterly from May to September, have important effects on water quality in Daya Bay.

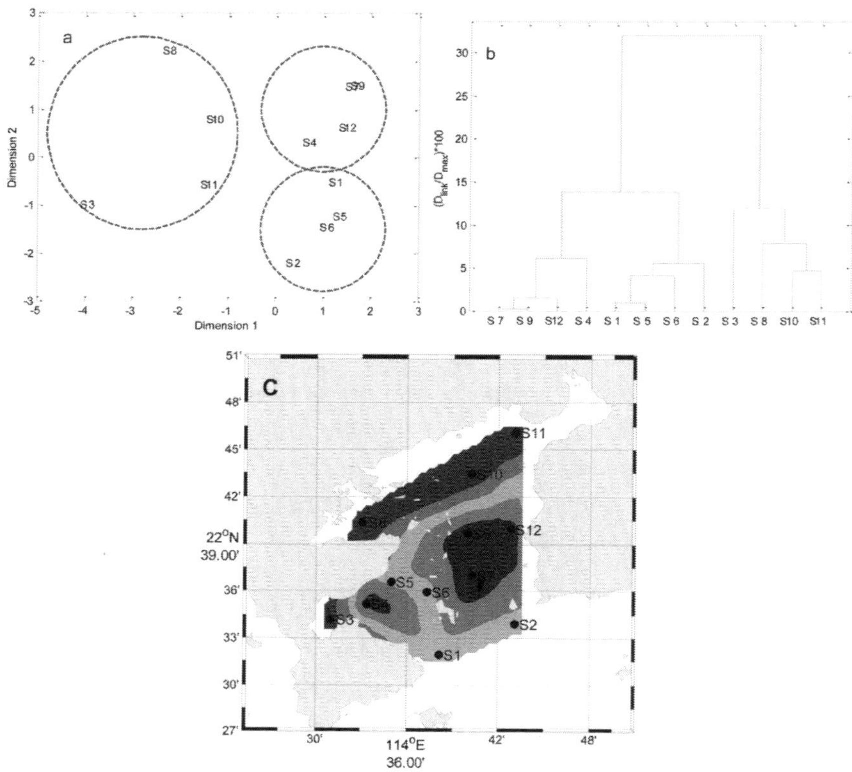


**Figure 9.** The results of multidimensional scaling analysis and cluster analysis: (**a**) Multidimensional scaling analysis plot for monitoring stations in overall spatial pattern; (**b**) Dendrogram based on Ward's method for monitoring stations in overall spatial pattern; (**c**) Map of the resulting zones of Daya Bay in overall spatial pattern.

# 5. CONCLUSIONS

In this study, the surface water quality in the Daya Bay was evaluated using the well known multivariate statistical methods PCA and MDS. Two factors explaining the 63.5257% of the total variance in the water quality data were identified. They were termed as nutrient factor and the biological factor. The nutrient factor explained 35.1039% and the agricultural use factor explained 28.4218% of the observed variance in the water quality data. The results of the PCA showed that urban wastewater and agricultural drainage waters were the main sources of the contamination in the bay. According to the Dimension 1, the monitoring stations (S3, S8, S10 and S11) were found to be the most dissimilar stations. These results may provide a basis for taking preventive action to reduce the pollution sources caused from aquiculture and human activities in the bay.

# ACKNOWLEDGMENTS

This research was supported by the project of knowledge innovation program of Chinese Academy of Sciences (No. KZCX2-YW-Q07-02, No. KSCX2-YW-Z-1024, No. SQ200913), the National Natural Science Foundation of China (No. 41076070), the Fund of Key Laboratory of Global Change and Marine-Atmospheric Chemistry, SOA (No.GCMAC0906), and Key Laboratory of Fishery Ecology and Environment, Guangdong Province (No.LFE-2010-14).

# REFERENCES

1. Xu, GZ. *Environments and Resources of Daya Bay*; Anhui Science Publishing House: He Fei, China, 1989.

2. Song, XY; Huang, LM; Zhang, JL; Huang, XP; Zhang, JB; Yin, JQ; Tan, YH; Liu, S. Variation of phytoplankton biomass and primary

production in Daya Bay during spring and summer. *Mar. Pollut. Bull* **2004**, *49*, 1036–1044.

3. Tang, DL; Kester, DR; Wang, ZD; Lian, JS; Kawamura, H. AVHRR satellite remote sensing and shipboard measurements of the thermal plume from the Daya Bay, nuclear power station, China. *Remote Sens. Environ* **2003**, *84*, 506–515.
4. Wang, CH; Qi, YZ; Li, JT; Xu, N; Chen, JF. Analysis and evaluation trophic status in aquaculture areas of Daya Bay.*Mar. Environ. Sci* **2004**, *23*, 25–28.
5. Qiu, YW; Wang, ZD; Zhu, LS. Variation trend of nutrient and chlorophyll contents and their effects on ecological environment in Daya Bay. *J. Oceanogr. Taiwan Strait* **2005**, *24*, 131–139.
6. Peng, YH; Sun, LH; Chen, HR; Wang, ZD. Study on eutrophication and change of nutrients in the Daya Bay. *Mar. Sci. Bull* **2002**, *21*, 44–49.
7. Wu, ML; Wang, YS. Using chemometrics to evaluate anthropogenic effects in Daya Bay, China. *Estuarine Coastal Shelf Sci* **2007**, *72*, 732–742.
8. Wang, YS; Lou, ZP; Sun, CC; Sun, S. Ecological environment changes in Daya Bay, China, from 1982 to 2004. *Mar. Pollut. Bull* **2008**, *56*, 1871–1879.
9. Wu, ML; Wang, YS; Sun, CC; Wang, HL; Dong, JD; Han, SH. Identification of anthropogenic effects and seasonality on water quality in Daya Bay, South China Sea. *J. Environ. Manage* **2009**, *90*, 3082–3090.
10. Wu, ML; Wang, YS; Sun, CC; Wang, HL; Dong, JD; Yin, JP; Han, SH. Identification of coastal water quality by statistical analysis methods in Daya Bay, South China Sea. *Mar. Pollut. Bull* **2010**, *60*, 852–860.

11. Wu, ML; Wang, YS; Sun, CC; Wang, HL; Lou, ZP; Dong, JD. Using chemometrics to identify water quality in Daya Bay, China. *Oceanologia* **2009**, *51*, 217–232.

12. Zhou, JL; Maskaoui, K. Distribution of polycyclic aromatic hydrocarbons in water and surface sediments from Daya Bay, China. *Environ. Pollut* **2003**, *121*, 269–281.

13. Yu, XJ; Yan, Y; Wang, WX. The distribution and speciation of trace metals in surface sediments from the Pearl River Estuary and the Daya Bay, Southern China. *Mar. Pollut. Bull* **2010**, *60*, 1364–1371.

14. Chen, TR; Yu, KF; Li, S; Price, GJ; Shi, Q; Wei, GJ. Heavy metal pollution recorded in Porites corals from Daya Bay, northern South China Sea. *Mar. Environ. Res* **2010**, *70*, 318–326.

15. Han, WY. *Marine Chemistry in South China Sea*; Science Publishing House: Beijing, China, 1998.

16. Huang, XN; Zhu, ZH; Xu, CM; Jin, QZ. Variation of Water Temperature in the Southwestern Daya Bay Before and After the Operation of Daya Bay Nuclear Power Plant. In *Annual Research Reports, Marine Biology Research Station at Daya Bay*; Pen, JP, Wang, ZD, Eds.; Science Publishing House: Beijing, China, 1998.

17. Ji, WD; Huang, SG. Relationship between Nutrient Variation in Daya Bay Water Column and Hydrographical and Biological Factors. In *Collections of Papers on Marine Ecology in the Daya Bay (II)*; Third Institute of Oceanography, SOA, Ed.; Ocean Publishing House: Beijing, China, 1990; pp. 123–132.

18. Wang, YS; Lou, ZP; Sun, CC; Wu, ML; Han, SH. Multivariate statistical analysis of water quality and phytoplankton characteristics in Daya Bay, China, from 1999 to 2002. *Oceanologia* **2006**, *48*, 193–211.

19. Zhan, B; Zeng, G; Li, L. Temperature and Salinity of Daya Bay. In *Collections of Papers on Marine Ecology in the Daya Bay (II)*; Third

Institute of Oceanography, SOA, Ed.; Ocean Publishing House: Beijing, China, 1990.

20. Chen, CC; Shiah, FK; Chung, SW; Liu, KK. Winter phytoplankton blooms in the shallow mixed layer of the South China Sea enhanced by upwelling. *J. Mar. Syst* **2006**, *59*, 97–110.

21. Wang, XP; Cai, WG; Lin, Q; Jia, XP; Zhou, GJ; Gan, JL; Lu, XY. The distribution variation of the nutrition salts in the waters of Daya Bay. *Trans. Oceanol. Limnol* **1996**, *4*, 20–27. [

22. Medina-Gomez, I; Herrera-Silveira, JA. Spatial characterization of water quality in a karstic coastal lagoon without anthropogenic disturbance: A multivariate approach. *Estuarine Coastal Shelf Sci* **2003**, *58*, 455–465. [

23. Huang, HH; Wang, ZD; Zhang, ZB; Pan, MX; Gao, HL; Peng, YH; Wei, GF; Zhu, ZH; Li, L. Preliminary study on the regional of chlorophyll—a and nutrients division of the distributions at Daya Bay in autumn. *Mar. Sci. Bull* **1999**, *18*, 32–38.

7

# FIRST TIME SERIES OF ESTIMATED HUMPBACK WHALE (*MEGAPTERA NOVAEANGLIAE*) ABUNDANCE IN PRINCE WILLIAM SOUND

**Suzanne F. Teerlink[1], Olga von Ziegesar[2], Janice M. Straley[3], Terrance J. Quinn II[1], Craig O. Matkin[3], Eva L. Saulitis[3]**

**1** Juneau Center, School of Fisheries and Ocean Sciences, University of Alaska Fairbanks, 17101 Point Lena Loop Road, Juneau, AK 99801, USA

**2** Eye of the Whale Research, P.O. Box 15191, Homer, AK 99603, USA

**3** Sitka Campus, University of Alaska Southeast, 1332 Seward Avenue, Sitka, AK 99835, USA

**4** North Gulf Oceanic Society, 3430 Main Street, Suite B1, Homer, AK 99603, USA

## ABSTRACT

In Prince William Sound (PWS), changes in abundance of humpback whales (*Megaptera novaeangliae*), one of PWS's primary marine predator species, have until now been largely unknown. Using a historical dataset (1978–2009), we constructed the first time series of estimated humpback whale abundance in western PWS that is also one of the longest time series used in analyses of humpback whale mark-recapture data. Photographs from this dataset were used to "mark" and re-sight individual animals using the unique pigmentation pattern on the ventral flukes of each whale in a mark-recapture analysis. Specifically, the POPAN implementation of the Jolly-Seber mark-recapture model in program MARK was used. Estimates of probabilities of capture and survival, recruitment parameters, and total abundance over the study were obtained, leading to a time series of abundance estimates. Our results show an increase from 39 (SE = 26) to 194 (SE = 17) whales (≈≈500 %) over the time series. The average annual rate of increase (ROI) was 4.53 % (95 % CI 3.28–5.79 %) which is only slightly lower than the 5–7 % ROI estimated for the North Pacific. Trends in the number of whales encountered per unit effort were not consistent with abundance estimates from mark-recapture, showing that sightability changes annually.

## KEYWORDS

Humpback whale, Jolly-Seber, MARK, Mark-recapture, *Megaptera novaeangliae,* Photo identification, POPAN, Rate of increase

## 1 INTRODUCTION

Humpback whale (*Megaptera novaeangliae*) populations in the North Pacific Ocean were severely overexploited by whaling and were reduced by over 90 % from an estimated pre-exploitation population size of 15,000 individuals (Johnson and Wolman 1984; Rice 1978). By the end of commercial whaling in the 1960's, this population was at an all-time low, estimated at 1,200 by Johnson and Wolman (1984), and 1,400 by Gambell (1976). Various protective measures were implemented to support recovery of this overexploited population, including: an International Whaling Commission (IWC) ban on humpback whale harvest in 1965, humpback whales were listed under the Mammal Protection Act (MMPA) of 1972, and listed under the Endangered Species Act (ESA) in 1973. Today, four decades later, humpback whales are still listed under the MMPA and ESA and are protected from commercial harvest by the IWC; however, abundance estimates of this species in the North Pacific Ocean are rare and often rely on limited data.

Since the end of large-scale industrial whaling, humpback whales in the North Pacific have been increasing in abundance. Using mark-resighting from 1990 to 1993, it was estimated that there were 10,000 individual humpback whales in the central and eastern North Pacific Ocean (Calambokidis et al. 1997). In 2004–2006 an international, collaborative research effort known as SPLASH (Structure of Populations, Levels of Abundance and Status of Humpbacks) was conducted throughout the entire North Pacific Basin (Calambokidis et al. 2008). As part of this study, population estimates were made using photographic identification in a mark-recapture analysis. They concluded that, at the end of their study in 2006, there were nearly 20,000 humpback whales in the North Pacific. Given a historic estimate of 15,000 individuals, the SPLASH abundance estimate suggests that there were more humpback whales in the North Pacific in 2006 than there had been prior to the onset of exploitation. In addition to estimating population size, SPLASH also calculated annual rate of increase (ROI) (Calambokidis et al. 2008). When they contrasted their estimates to that of 1993 of approximately 10,000 in the North Pacific (Calambokidis et al. 1997), they found a 4.9 % ROI over the 13 year period. However, when they instead compared against the 1966 population estimate of 1,400 (Gambell 1976), they found a 6.8 % ROI over the 39 year period. Given these results, SPLASH concluded that humpback whale abundance in the North Pacific is increasing at 5–7 % annually. In a different study, the maximum theoretical biological ROI was estimated for humpback whales using best-case scenario values for adult survival, young-of-the-year survival, pregnancy rates, and age at first reproduction. Here, it was found that the maximum plausible ROI due to biological reproduction is 11.8 % (Zerbini et al. 2010). While these studies arrived at different estimates of ROI, it still can be concluded that the humpback whale population of the North Pacific Ocean is increasing and may have reached or exceeded pre-exploitation abundance levels.

Humpback whales are migratory; in the North Pacific Ocean they generally spend summers in northern latitudes feeding and winters in tropical waters breeding and calving (Johnson and Wolman 1984).

While on the feeding grounds, humpback whales use their baleen to filter small schooling forage fish and invertebrates, such as Pacific herring (*Clupea pallasii*), capelin (*Mallotus villosus*), the juvenile stages of Pacific salmon (*Oncorhynchus* spp.), and euphausiids (Nemoto 1970; Straley 2009, pers. comm.). Based on energetic analyses it was estimated that humpback whales consume approximately 0.4 tons of prey each day while on the feeding grounds (Witteveen et al. 2006). While humpback whales can be found feeding in Alaska every month of the year, their abundance peaks in summer and fall months and they are least abundant in the late winter and spring (Baker et al. 1992; Rice et al. 2011; Straley 1990, 1994).

One important feeding ground for humpback whales in the North Pacific is Prince William Sound (PWS). At present, very little information exists on the current and historic abundances of humpback whales in PWS. Only a few historic point estimates of humpback whales in PWS have been made including 84 (SE == 11) individuals in 1980–1983 and 100 (SE == 20) individuals in 1983–1984, both found using a Chapman mark-recapture estimator (Matkin and von Ziegesar 1985) and an estimate of 19 individuals in 1985 using a Schnabel estimator in a low effort year with few whales encountered and identified (von Ziegesar and Matkin1986). From these estimates, the population in PWS was estimated to increase at 6 % annually, which is in the range of estimates for the entire North Pacific Basin determined by SPLASH (von Ziegesar and Matkin1986). No other estimates of abundance were made until 2007–2009, when a field study was undertaken to investigate the magnitude and composition of prey consumption by humpback whales in PWS (Rice et al.2011). Using mark-recapture methods, late-season abundance of humpback whales was estimated at 135 (SE == 12) whales in September-March of 2009 (Rice et al. 2011).

In this paper we use a comprehensive dataset of 30 years, from 1978 to 2009, of humpback whale surveys in PWS from Eye of the Whale Research (EOW) and North Gulf Oceanic Society (NGOS) to examine trends in abundance and develop estimates of humpback whale population size. This dataset has been used to catalog individual humpback whales in the area and for various other analyses (von Ziegesar and Matkin1985; von Ziegesar 1992; von Ziegesar et al. 2001; von Ziegesar 2010, 2012); however, no recent estimates of humpback whale population size or population change over time have been generated from these data.

This project uses this dataset to develop the first time series of humpback whale abundance in western PWS. Specifically, we employ sightings per unit of effort and mark-recapture to evaluate relative and absolute abundance, respectively. Estimates are then used to characterize the trends in abundance for this area and compare them to the trends documented for the entire North Pacific Basin over the same time period. Given that such long datasets of mark-recapture data are rarely available, this presents an opportunity for evaluation of the performance of a commonly used open population model. To our knowledge, this is the longest time series used in any mark-recapture analysis of humpback whale abundance.

# 2 METHODS

## 2.1 Study area

Prince William Sound is located in the Gulf of Alaska (60∘35′N,147∘10′W60∘35′N,147∘10′W), and is comprised of numerous islands, glacial fjords, and bays. The western portion of PWS was designated as our study area (Fig. 1), as this is where the majority of surveys from the dataset were conducted. The dataset also includes some data from neighboring areas; however, because these areas are not consistently surveyed over the time series, these data were removed from our analyses.

## 2.2 Effort

Three measures of effort were available from the dataset: nautical miles traveled, hours spent searching, and number of boat-days. For early years, where track lines were recorded by hand on photocopied maps, track lines were manually digitized. In more recent years, many of the surveys had digital track lines from handheld GPS units. For surveys conducted within the defined study area, each measure of effort (nautical miles, hours, and days) was totaled by year, and the different measures were analyzed to determine which was most appropriate for this study (Table 1). Many surveys had no record of the track line taken and/or the number of hours spent searching; years known to have >>10 % missing effort data are indicated in Table 1.

**Figure 1.** Map of Prince William Sound, Alaska, showing the designated study area of western Prince William Sound (*circled*)

**Table 1.** Days, nautical miles (nm), and hours of effort in western Prince William Sound by year

| Year | Days | nm | Hours | Count | Whales ID'ed | New IDs |
|---|---|---|---|---|---|---|
| 1978 | 11 | N/A | N/A | 18 | 12 | 12 |
| 1979 | 8 | N/A | N/A | 23 | 16 | 13 |
| 1980 | 36 | N/A | N/A | 181 | 26 | 1 |
| 1981 | – | – | – | – | – | – |
| 1982 | – | – | – | – | – | – |
| 1983 | 29 | N/A | N/A | 93 | 36 | 17 |
| 1984 | 135 | 6,124 | N/A | 78 | 38 | 18 |
| 1985 | 65 | 2,404 | 165 | 39 | 15 | 4 |
| 1986 | 55 | 2,527 | 223 | 55 | 27 | 7 |
| 1987 | 29 | 1,111 | 154 | 59 | 12 | 1 |
| 1988 | 73 | 2,331 | 286 | 200 | 40 | 15 |
| 1989 | 215 | 8,738 | 718 | 340 | 60 | 24 |
| 1990 | 252 | 10,585 | 1,172 | 595 | 70 | 19 |
| 1991 | 187 | 8,451 | 670 | 252 | 63 | 14 |
| 1992 | 134 | 5,665 | 684 | 316 | 86 | 26 |
| 1993 | 72 | 3,019 | 56* | 82 | 46 | 13 |
| 1994 | 83 | 3,413 | 136* | 189 | 62 | 12 |
| 1995 | 125 | 5,976 | 123* | 132 | 49 | 10 |
| 1996 | 92 | 4,158 | 111* | 298 | 69 | 21 |
| 1997 | 126 | 5,722 | 194* | 262 | 83 | 27 |
| 1998 | 98 | 4,451 | 159* | 157 | 63 | 11 |
| 1999 | 98 | 4,514 | 120* | 121 | 39 | 8 |
| 2000 | 83 | 4,001 | 195* | 188 | 59 | 10 |
| 2001 | 87 | 4,272 | 149* | 197 | 79 | 11 |
| 2002 | 107 | 5,226 | 162 | 186 | 73 | 15 |
| 2003 | 79 | 3,722* | 67* | 94 | 41 | 6 |
| 2004 | 142 | 7,785 | 61* | 122 | 51 | 5 |
| 2005 | 112 | 4,686* | 81* | 391 | 73 | 6 |
| 2006 | 61 | 2,696* | 58* | 148 | 58 | 4 |
| 2007 | 71 | 4,013* | 42* | 180 | 67 | 16 |
| 2008 | 54 | 814* | 15* | 176 | 56 | 8 |
| 2009 | 64 | 473* | 74* | 218 | 104 | 36 |

An asterisk indicates that the effort measure is unavailable for >10% of the surveys. This table also shows total whales counted throughout the survey season (Count), total number of unique whales photo identified in that year (Whales ID'ed) and of those, the number of whales identified that were new to the study (New IDs)

[a] No sampling occurred in 1981–1982

## 2.3 Description of available data

The dataset includes humpback whale sightings and identification photos of the ventral surface of humpback whale flukes, taken annually between April and September from 1978 to 2009. The documentation of the unique pigmentation patterns of the ventral flukes is a commonly used method to identify and track individual humpback whales over time (Katona et al. 1979). No data were collected in 1982, and data from 1981 was thrown out because effort in this year was relatively small and data collection was compromised by weather complications. Photos were taken using single-lens reflex cameras from 1978 to 2002 and using digital cameras from 2003 to 2009. Photo identification records include >>5,000 photographic sightings of ~575~575 individual humpback whale flukes over the time series. Depending on the year, between one and three research teams collected data on separate boats. Research teams however had different specific goals; namely, some specifically targeted humpback whales (von Ziegasar, EOW), while others targeted killer whales (*Orcinus orca*) and gathered humpback whale data only opportunistically (Matkin and Saulitis, NGOS). The total number of whales sighted, unique identifications, and newly identified individuals are summarized by year (Table 1).

## 2.4 Data organization

Data describing effort, humpback whale sightings, and individual characteristics in Microsoft Excel files (EOW) were transferred into the Alaska Humpback Whale Database, a Microsoft Access database designed specifically for storage and querying of humpback whale photographic data (developed by co-author Straley and colleagues). The database contains tables to hold information pertaining to a given whale at the time of the sighting, unique and long-lasting characteristics of that whale, and information specific to the survey effort. During this entry process, sighting histories were error-checked by verifying Excel entries with actual photos and with hard-copy datasheets. The best photograph of each individual whale each year was identified and, for years where photos were archived as negatives (1987–2002), the photos were digitized by scanning the negatives.

## 2.5 Humpback whale time series: count per unit effort

All recorded sightings of humpback whales—both those whales that were photographed and those that were sighted and noted only—were tallied for each year. An annual index of relative abundance was calculated as the number of sightings divided by the amount of effort. A quadratic regression line was fitted to determine if there was a linear or curvilinear trend over time. Additionally, an index of the number of unique whales identified each year divided by the amount of effort was also constructed. Again, a quadratic regression line was fitted to assess any trend over time.

## 2.6 Humpback whale time series: mark-recapture modeling

### 2.6.1 Mark recapture modeling consideration: photo quality

Quality of identification photos was considered for the mark-recapture portion of this analysis, as poor quality photos can cause misidentification of individuals and result in bias in a mark-recapture setting. Typically, in the case of photo identification of humpback whales, it is unlikely to have false positive errors; namely, instances where photos are wrongly designated as re-sightings. However, the occurrence of false negative errors-instances where photos are wrongly designated as new sightings when they are actually resightings-increases with the inclusion of poor quality photos (Friday et al. 2000, 2008; Stevick et al. 2001). Efforts were taken to reduce the occurrence of this type of error by removing the photographs determined to be of poor quality. This was done by ranking the quality of the photo with a value of 1 through 3 in six categories. Values from each of the six categories were then summed to generate an overall index of quality for each photo; thus, index values could range from 6 to 18, with lower values indicating a better quality photo. Criteria used to rank these photos are a modified version of the protocols described by others1(Calambokidis et al. 2008; Friday et al. 2000). We deemed all photos with a total photo quality index greater than or equal to 9 to be "poor" and removed them from the analysis.

## 2.7 Mark recapture modeling consideration: fluke distinctiveness

Fluke distinctiveness could also play a role in producing false negative errors, instances where photos are wrongly designated as new sightings, which would introduce positive bias in abundance in a mark-recapture analysis (Friday et al. 2000, 2008; Stevick et al. 2001). Flukes bear wildly different patterns and markings, making some individuals more identifiable than others. Thus, it is more likely that a match of an individual with nondescript flukes would be overlooked resulting in a false negative error. To evaluate the importance of fluke distinctiveness in the sighting histories of this dataset, we evaluated all flukes in this analysis (4306 sightings of 356 individual whales) by ranking them with four criteria. The ranking system used2 is a modified version of that by Friday et al. (2000). Each pair of flukes was assigned a value, 1 through 3, in each category. Values from each of the four categories were then summed to generate an overall index of fluke distinctiveness for each pair of flukes; index values could therefore range from 4 to 12, with lower values indicating more distinct flukes. This index was then used to measure the importance of fluke distinctiveness in recapture probability (after the photo quality standard was applied) by assessing the relationship, if any, between total number of sightings for each individual and the distinctiveness index of its flukes. Additionally, this index was used as an individual covariate to the probability of capture in mark recapture models as explained in later sections.

### 2.7.1 Mark recapture modeling

For this study, we chose only open-population models, because population-level changes of humpback whales over many years result from recruitment (births or immigration) and loss (death and emigration). The mark-recapture program MARK was used to obtain maximum likelihood parameter estimates (MLE's) and to compare various models using the POPAN implementation of the Jolly-Seber model (Schwarz and Arnason1996; White and Burnham 1999). Like most commonly-used mark-recapture models, the Jolly-Seber model generates estimates of parameters by modeling sighting histories as a multinomial distribution. Its structure allows several model configurations to be estimated, permits the inclusion of covariates, and allows for parameters to be added, fixed, and set equal across time periods, groups, and strata. Sightings of all first year calves were removed from the mark-recapture analyses as their probability of capture is complicated by co-occurrence with their mothers.

The POPAN procedure in program MARK runs a variety of the widely used and accepted models for open populations (Schwarz and Arnason 1996; Schwarz 2001) and was chosen because it generates annual estimates of abundance (N^t)(N^t) to form a time series. This model is a modified version of the traditional open Jolly-Seber model and, based on the sighting histories, estimates survival, ϕtϕt, and probability of capture, ptpt, "births" or net number of new entrants into the population (*pent*) in year tt, and N^N^, the overall population size. Estimates of standard error come from the variance-covariance matrix, and the estimates of the lower and upper 95 % confidence intervals of the ϕϕ's andpp's are made using the likelihood profile for the MLE (Lebreton et al. 1992).

We also examined a commonly-used alternative model from Pradel (1996), because of the addition of a population growth rate parameter. However, we limit our reporting on the Pradel model, because its estimates are very similar to those of the Jolly-Seber model yet less precise.

The assumptions for the Jolly-Seber model are as follows Seber (1982):

1. Each animal in the population has the same probability of capture, ptpt, at sampling occasion tt.
2. Each animal has the same probability of survival, ϕtϕt, at a given interval between sampling occasions ttand t+1t+1.

3. Marks are not lost or misidentified, and all marks are reported.

4. Sampling is instantaneous, relative to the interval between sampling occasions.

Various Jolly-Seber model configurations were developed within MARK and compared to identify the best model. The ϕϕ parameters were modeled as dependent on time period, denoted ϕ(t)ϕ(t); or constant, ϕ(.)ϕ(.). The pp parameters were modeled with the following scenarios: time-dependent, p(t)p(t); constant, p(.)p(.); a linear function of effort, pp(eff); dependent on fluke distinctiveness through use of an individual covariate, pp(FD); and dependent on a combination of both time and fluke distinctiveness, p(t,FD)p(t,FD). It was expected that there would be variability in capture, but unknown if relative effort or the more flexible time effects would best explain this variability. Further, fluke distinctiveness was suspected to affect sightability (see Sect. 4.3.1). Therefore, models with all possible combinations of ϕϕ and pp configurations were constructed, because no combination could be ruled out in terms of biological plausibility. The parameter for the probability of new recruits (entrants), *pent*, was modeled dependent on time in all cases because it is well known that the number of births varies from year to year. The overall population estimate, N^N^, was estimated, which is the superpopulation size or the number of whales available at some point throughout the duration of the study. From N^N^, annual estimates of abundance, N^tN^t were derived within the MARK software.

Competing models were evaluated using the Akaike's Information Criterion (AIC) adjusted for small sample sizes, AICc (Burnham and Anderson 2002, pg 128). The best model was selected using the guidelines offered by Burnham and Anderson (2002), pg 128: the model with the lowest AICc value is considered "best" and other models are given in terms of the difference, ΔΔAICc with the best model. Any model with ΔΔAICc ≤≤ 2 has "...no credible evidence that the model should be ruled out". Models with ΔΔAICc of 2–4 offer "weak evidence of not being the best model", models with ΔΔAICc of 4–7 offer "definite evidence", ΔΔAICc of 7–10 offer "strong evidence", and ΔΔAICc >>10 offer "very strong evidence of not being the best model". Goodness of fit was examined through deviance residuals and no unwanted patterns were seen.

### 2.7.2 Sensitivity analysis: early year photo quality

Photos from some early years were not available to be assessed for quality (specifically: 1980, 1983–1985, and 1987). Because there is no means of measuring or compensating for this, we anticipated that there would be some level of bias introduced by data from these years. To address this, we conducted a sensitivity analysis to determine the effect of these early years on the remainder of the time series. We did this by re-running the selected Jolly-Seber model using only data from 1988 to 2009, years where all photos were assessed for quality. Estimates from this model were then compared to estimates from the complete time series both graphically and by calculating the correlation coefficient for the overlapping portion of both time series.

### 2.7.3 Sensitivity analysis: survival

To assess the impact of emigrating whales on the abundance estimates, another sensitivity analysis was conducted. Survival probability fixed equal to 0.98 in the selected model and the other parameters were re-estimated. A survival probability of 0.98 was selected because this is the rate that was found in the only other assessment of survival of humpback whales in PWS (Mizroch et al. 2004). The results were then compared to those from the base Jolly-Seber model to assess how a different survival would affect the model outputs.

## 2.8 Comparison of time series approaches

Abundance time series results from all four time series models, the two count per effort models and the abundance estimates from Jolly-Seber and Pradel mark-recapture models, were compared in a correlation analysis to evaluate their similarity for this dataset.

## 2.9 Rate of increase

The ROI, frequently used in demography, is defined as the geometric mean of the average annual population growth rate minus 1 (expressed as a percentage rate of increase). It is calculated in Eq. (1) from the changes in logarithms of abundance estimates $\{\Delta_y = \Delta \ln(\widehat{N}_y)\}$ produced by the Jolly-Seber model

$$ROI = \exp\left(\frac{1}{Y}\sum \Delta \ln\left(\widehat{N}_y\right)\right) - 1 = \exp\left(\overline{\Delta}\right) - 1, \tag{1}$$

in which $\overline{\Delta} = \sum \Delta_y / Y$ and Y is the number of years. The approximate SE of ROI (equal to the SE of ROI + 1) was found from the delta method (Seber 1982, pp. 7–9). It is calculated as in Eq. (2)

$$SE(ROI) = \overline{\Delta} SE\left(\overline{\Delta}\right) = (ROI + 1)\, SE\left(\overline{\Delta}\right), \tag{2}$$

for which the standard error SE(Δ¯¯¯)SE(Δ¯) is found empirically in Eq. (3):

$$SE\left(\overline{\Delta}\right) = \left(\frac{1}{Y}\right)\sqrt{\sum SE^2(\Delta_y)}, \tag{3}$$

from the estimates {SE(Δy)}{SE(Δy)} produced by the MARK software. This calculation ignores estimated covariances between the {Δy} values, but these were small compared to the variances of N^N^. Equation (4) was used to approximate a 95 % confidence interval for ROI.

$$LCI(ROI), UCI(ROI) = ROI \pm 1.96 SE(ROI) \tag{4}$$

This method is the same used by Calambokidis et al. (2008) to generate ROI estimates for the SPLASH study, and was employed to make direct comparisons possible.

# 3 RESULTS

## 3.1 Effort

From examination of effort data (Table 1), our preferred effort measure for this study is in days, as it was available consistently for all years. Furthermore, effort in days is comparable to effort in nautical miles, which normally would be thought to be a more accurate measure.[3] In years where nautical miles traveled were consistently recorded, 1984–2002 and 2004, nautical miles and number of days on the water were significantly correlated (Pearson $r=0.99,p<0.001r=0.99,p<0.001$).

## 3.2 Humpback whale time series: count per unit effort

Counts of humpback whale sightings from this study ranged from 0.58 whales per day in 1984 to 5.03 whales per day in 1980 (Fig. 2). There was no significant trend in this time series ($p=0.19p=0.19$). Number of unique individuals identified each year divided by days of effort ranged from 0.23 in 1985 to 2.00 in 1979 with a trend that was similar to the average rate.[4] As in the sightings per effort time series, there was no significant trend in the unique individual per effort time series ($p=0.48p=0.48$).

## 3.3 Humpback whale time series: mark-recapture modeling

### 3.3.1 Mark recapture modeling consideration: fluke distinctiveness

The number of sightings of all individuals as a function of their fluke distinctiveness index had a slight, yet significantly increasing trend ($p=0.015p=0.015$, Fig. 3). This indicates that some level of heterogeneity bias may be present in the dataset based on recognizability of flukes.

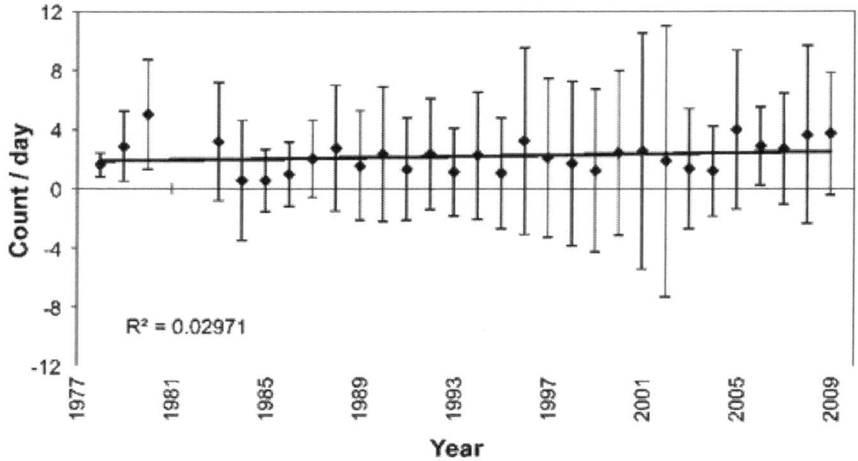


**Figure 2.** Average humpback whale count per day of effort for each year; all whales seen and recorded were included in the count, regardless of whether or not the whale was photographically identified. *Error bars* are 1 standard deviation. Trend line is a linear regression

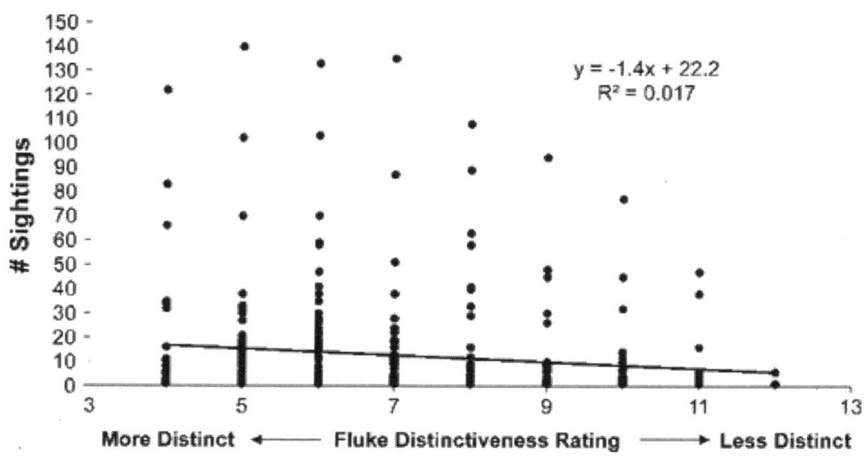


**Figure 3.** Number of sightings of each individual showed plotted by fluke distinctiveness index value

### 3.3.2 Mark recapture modeling

Models including $\phi(t)\phi(t)$ were unrealistic due to parameter estimates converging too close to the boundary of 1. This is likely indicative of too sparse data and the difficulty in detecting survival for a species with such high survival rates; results from these models are not presented. Model $\phi(.)$,p(t),pent(t),N$\phi(.)$,p(t),pent(t),N was selected as the best Jolly-Seber model (Table 2). This model has constant survival, $\phi\phi$, over the time series, estimated at 0.94 (95 % CI 0.93–0.95), and time dependent probability of capture, pp, with estimates ranging from 0.16 to 0.57 (Table 3), and no obvious trend over years (Fig. 4). Further, there were no issues with parameters converging on the boundary in this model, so there was no need to fix parameters equal and we had separate estimated parameters for pp in each year of the time series. The mark-recapture model with an effort covariate had a significantly poorer fit with ΔΔAIC much greater than the criterion of 10. Estimates of pp, did, however, show a significant linear relationship (p=0.034p=0.034) with effort (Fig. 5), which was not improved by adding a quadratic term. The addition of fluke distinctiveness index values as an individual covariate did not improve the model fit (Table 2). The general trend in abundance (N^t)(N^t) was quadratic and both variables in the polynomial were significant (Year, p=0.0030, Year2,p=0.0031p=0.0030,Year2,p=0.0031; Fig. 6). The overall abundance estimate, N^N^, was 396 (95 % CI 371–431). Estimates of the *pent* parameter are not reported in this paper, but are available from the senior author.

### 3.3.3 Sensitivity analysis: early year photo quality

Abundance estimates from the same Jolly-Seber model, $\phi\phi(.)$ p(t),pent(t),Np(t),pent(t),N that was re-run using only years with complete assessments of photo quality (1988–2009), showed a statistically significant correlation between paired points in overlapping years of 0.99 (p<0.001p<0.001; Fig. 7). This high degree of correlation indicates that photographs from early years were not introducing bias in estimates of later years.

**Table 2.** Summary of Jolly-Seber mark-recapture model fits, analyzing sighting histories from 1978–2009, based on survival (ϕϕ), and probability of capture (p)(p)

| Jolly-Seber model | K | AICc | Δ AICc |
|---|---|---|---|
| $\phi(.)p(t)\ pent(t)\ N$ | 56 | 4202.2 | – |
| $\phi(.)p(t, \mathrm{FD})\ pent(t)\ N$ | 57 | 4202.7 | 0.5 |
| $\phi(.)p(\mathrm{eff})\ pent(t)\ N$ | 27 | 4233.2 | 31.0 |
| $\phi(.)p(\mathrm{FD})\ pent(t)\ N$ | 25 | 4243.8 | 41.6 |
| $\phi(.)p(.)\ pent(t)\ N$ | 23 | 4244.4 | 42.2 |

The number of parameters (K), the corrected Akaike Information Criterion (AICc), and the difference in AICc compared to the best fit model (Δ AICc) are found for each model described in this table. Models with time dependent $\phi$ are not included in this table because many $\phi_t$ parameters converged too close to the boundary of 1, and therefore did not produce realistic results. All models have time dependent parameters for probability of new entrants (*pent*) and an overall estimate of abundance (*N*)

**Table 3.** Estimates and corresponding estimated standard error (SE) for probability of capture, p^p^, and the derived estimates of abundance, N^N^, by year in the selected Jolly-Seber model, ϕ(.),p(t),pent(t),Nϕ(.),p(t),pent(t),N

| Year | $\hat{p}$ Estimate | $\hat{p}$ SE | $\hat{N}$ (Derived) | $\hat{N}$ SE |
|---|---|---|---|---|
| 1978 | – | – | 38.5 | 26.8 |
| 1979 | 0.23 | 0.14 | 51.3 | 29.5 |
| 1980 | 0.49 | 0.12 | 52.0 | 12.2 |
| 1983 | 0.57 | 0.09 | 52.8 | 8.6 |
| 1984 | 0.44 | 0.08 | 73.3 | 11.3 |
| 1985 | 0.20 | 0.05 | 71.1 | 12.2 |
| 1986 | 0.37 | 0.06 | 66.9 | 8.1 |
| 1987 | 0.16 | 0.05 | 62.7 | 7.7 |
| 1988 | 0.28 | 0.06 | 84.3 | 13.7 |
| 1989 | 0.46 | 0.06 | 95.9 | 11.5 |
| 1990 | 0.49 | 0.06 | 96.7 | 9.7 |
| 1991 | 0.52 | 0.05 | 91.6 | 8.3 |
| 1992 | 0.56 | 0.05 | 117.0 | 10.3 |
| 1993 | 0.27 | 0.04 | 124.6 | 13.3 |
| 1994 | 0.44 | 0.05 | 120.9 | 10.2 |
| 1995 | 0.33 | 0.05 | 125.0 | 11.4 |
| 1996 | 0.38 | 0.05 | 156.9 | 14.7 |
| 1997 | 0.41 | 0.05 | 154.1 | 12.4 |
| 1998 | 0.36 | 0.04 | 147.9 | 11.4 |
| 1999 | 0.25 | 0.04 | 148.8 | 10.2 |
| 2000 | 0.38 | 0.04 | 139.5 | 9.9 |
| 2001 | 0.47 | 0.05 | 141.8 | 10.3 |
| 2002 | 0.40 | 0.04 | 144.2 | 10.0 |
| 2003 | 0.26 | 0.04 | 135.2 | 9.9 |
| 2004 | 0.30 | 0.04 | 131.7 | 10.8 |
| 2005 | 0.50 | 0.05 | 124.7 | 9.6 |
| 2006 | 0.46 | 0.05 | 118.7 | 9.4 |
| 2007 | 0.39 | 0.05 | 143.3 | 11.4 |
| 2008 | 0.31 | 0.04 | 134.3 | 11.2 |
| 2009 | 0.47 | 0.05 | 193.9 | 17.4 |

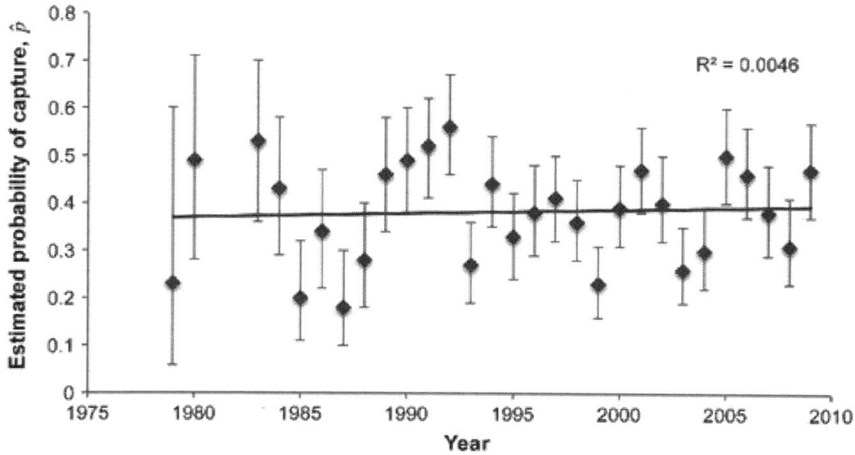


**Figure 4.** Capture probability parameter estimates and corresponding approximate 95 % confidence intervals the parameters in the selected Jolly-Seber model, ϕ(.),p(t),pent(t),Nϕ(.),p(t),pent(t),N

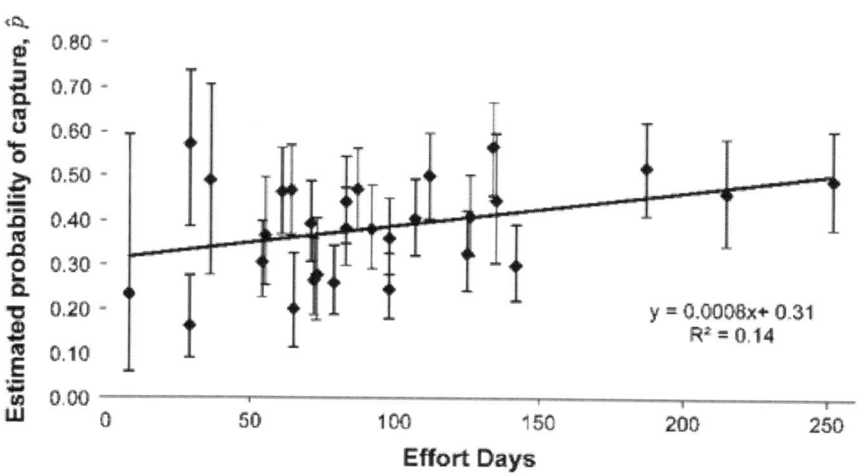


**Figure 5.** Estimates of probability of capture, pp, from the selected Jolly-Seber model versus the number of effort days.*Error bars* indicate approximate 95 % confidence intervals

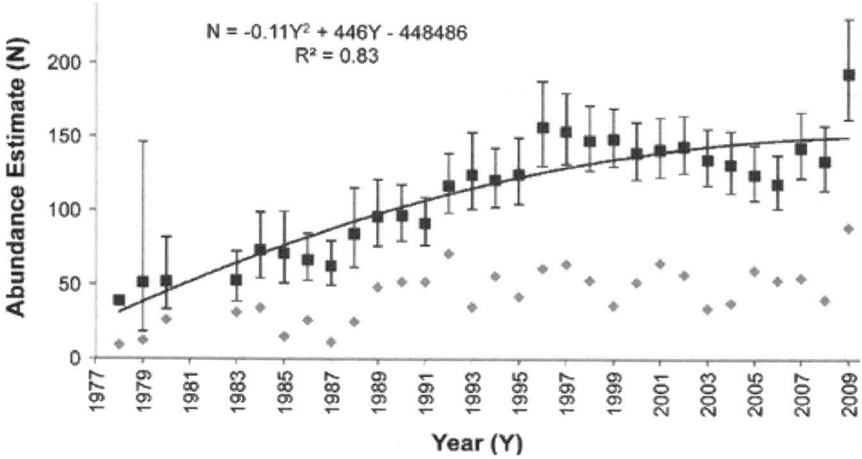


**Figure 6.** Abundance estimates from the selected Jolly-Seber model (*black squares*) by year. The number of unique individuals for each year is shown in *gray diamonds*. No data were available for 1981 or 1982. *Error bars*indicate the approximate 95 % confidence intervals

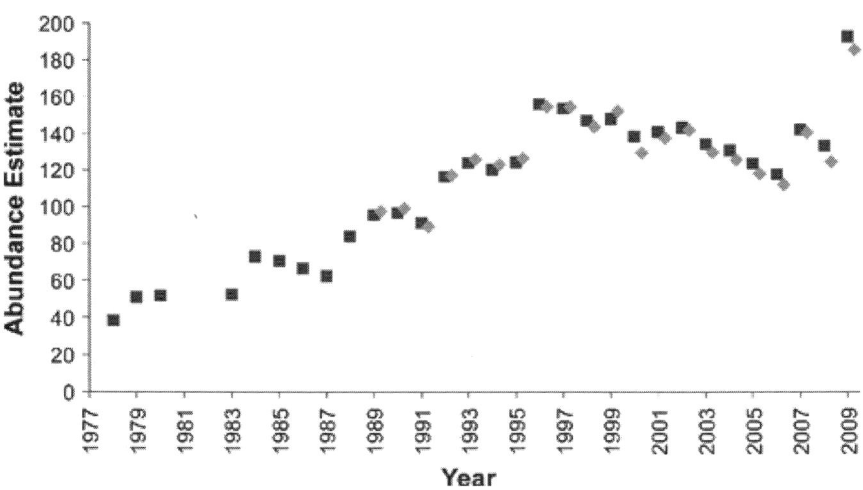


**Figure 7.**Abundance estimates of humpback whales in PWS from Jolly-Seber models using all available data, 1978–2009 (*black squares*), or using only 1988–2009 data (*gray diamonds*). The *gray diamonds* are offset to make them separate from the *black squares*. No data were available for 1981 and 1982

### 3.3.4 Sensitivity analysis: survival

The Jolly-Seber model, ϕϕ(.), p(t),pent(t),Np(t),pent(t),N, was rerun with a pre-defined constant survival (making the model, ϕ=0.98,p(t),pent(t),Nϕ=0.98,p(t),pent(t),N; Fig. 8). There was a high correlation between the paired annual abundance estimate points of these model results and the original Jolly-Seber model, however, it had a much poorer fit (ΔΔ AICc = 114.6).

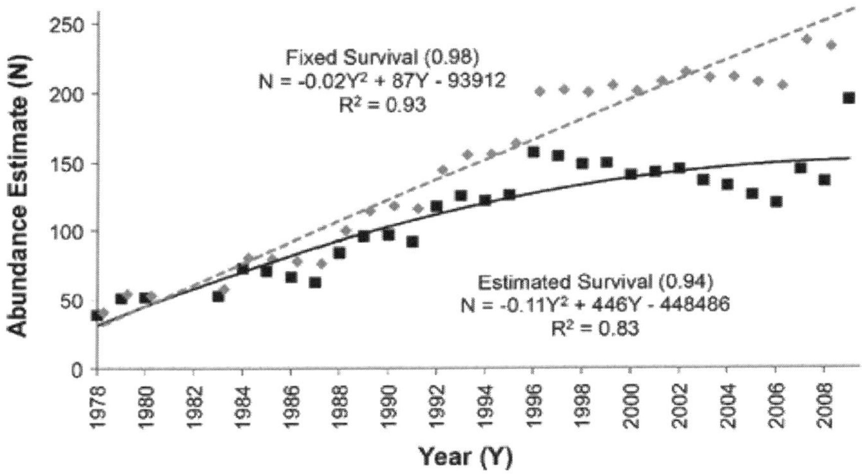


**Figure 8.**Abundance of humpback whales in Prince William Sound from Jolly-Seber models where survival is estimated (*black squares*) and where survival is fixed to 0.98 (*gray diamonds*). The *gray diamonds* are offset to make them separate from the *black squares*. Polynomial trend lines for both model results are displayed on the graph; the black solid line shows the trend for the model with estimated survival and the *gray dashed line* shows the trend for the model with fixed survival

## 3.4 Comparison of time series approaches

Neither of the count per effort time series was correlated with either the Jolly-Seber or Pradel mark-recapture time series5 (Table 4). The count per effort and the number of unique IDs per effort time series had a small amount of correlation. The time series from the two mark-recapture approaches were highly correlated ($p<0.001p<0.001$), showing that they produced extremely similar trends (Table 4).

## 3.5 Rate of increase

The ROI estimate from the Jolly-Seber model results was 4.53 % (95 % CI 3.28–5.79 %). This estimate of ROI is comparable to the SPLASH ROI estimate (Calambokidis et al. 2008) with UCI overlapping with the lower end of the SPLASH estimate of 5–7 %. There was no significant change in the individual annual rate of population change over time ($p=0.68p=0.68$). Our estimates are feasible as they are well within the estimated maximum theoretical biological ROI of 11.8 % per year (Zerbini et al. 2010).

# 4 DISCUSSION

## 4.1 Effort

Although we found the number of survey days to be the best effort measure for this analysis, a finer scale would have been preferable because the time and distance covered in a given survey day are expected to be variable. However, due to missing data values, the number of survey days per year was the only complete measure of effort available over the complete time period. In addition, we were able to show that this measure of effort was a reasonable proxy by demonstrating that it was highly correlated with nautical miles for those years where nautical miles was available. We attempted to include effort as a covariate to probability of sighting as this would have decreased the overall number of parameters in the model; however, effort data was not related to probability of sighting in the mark-recapture models. Therefore, we did not select models that included effort covariates for this dataset.

**Table 4.** Correlation matrix comparing the different time series models

| | Mark-recapture models | | Count per effort | Unique IDs per effort |
|---|---|---|---|---|
| | Jolly-Seber abundance estimate | Pradel abundance index | | |
| Jolly-Seber abundance estimate | 1 | 0 | 0 | 0 |
| Pradel abundance index | **0.939** | 1 | 0 | 0 |
| Count per effort | −0.001 | 0.096 | 1 | 0 |
| Unique IDs per effort | 0.089 | 0.182 | **0.623** | 1 |

Correlation values that are shown in bold are significant ($p < 0.05$)

## 4.2 Humpback whale time series: count per unit effort

Count per unit effort time series, in the form of number of sightings and number of unique individuals sighted per unit of effort, did not appear to provide reliable information about population trends for this dataset. Based on local knowledge, and known broader scale population trends (specifically, the North Pacific wide study, SPLASH), we would have expected to see a generally increasing trend in abundance over time. That we did not is likely due to a saturation effect inherent in the data collection procedures. Each whale seen in a survey requires an appreciable dedication of time in order to successfully photograph the ventral flukes of the animal for identification. This generally limits a researcher to a maximum of 10–20 individual whale encounters in a day. So, even if whale densities were increasing, it would not be possible to detect these changes beyond a certain density.

Another contributing feature to the implausibility of the count per effort series is that field sightings are complicated by many factors aside from the actual time or length of the survey in our study, including: weather conditions, sightability, prior knowledge of humpback whale locations, logistical issues, and some research teams gathering humpback whale data opportunistically only. Without the addition of more personnel, any combination of these factors would make the traditional time and distance measurements less reliable. To avoid these issues it would be necessary to design a rigorous study with predefined transects, weather and sightability constraints, and consistent on- and off-effort time and distance measurements.

The large variability in estimates is likely reflective of the nature of the survey logistics. First, the complex geography of the study area makes surveys more challenging as whales can be found deep in bays and inlets that are not observable from the main water channels. Secondly, the study area is larger than could be covered in a single day's survey, and because humpback whales are not found homogeneously throughout the study area, some survey days had relatively high whale counts, whereas others yielded zero whale counts.

## 4.3 Humpback whale time series: mark-recapture modeling

### 4.3.1 Mark recapture modeling consideration: fluke distinctiveness

We know from the relationship between sighting frequency and the fluke distinctiveness index that there is some bias toward more distinct flukes in the sighting histories. There are two possible mechanisms in the photo identification process that could have contributed to this bias: (1) removal by photo quality cutoff and (2) misidentification errors. Despite attempts to evaluate photo quality and distinctiveness of flukes independently, it is possible that more distinct flukes were more likely to meet the photo quality criteria than less distinct flukes since these measures are inherently subjective and rankers easily mistake recognizability with photo quality. In a mark-recapture setting, this bias works to give more distinct flukes a higher probability of capture and thus is a violation of the homogeneity assumption for capture probability of all animals. The over-dispersion parameter, c^c^, is estimated in program MARK, but we were not confident in its reliability because there are many capture histories with low frequencies of occurrence. Therefore we did not attempt to correct for over-dispersion in this manner. Further, heterogeneity usually creates a negative bias, which is not of much concern here.

Alternatively, the relationship between sighting frequencies and the fluke distinctiveness index could occur by false negative misidentification errors, or instances where matches are overlooked. This could be caused by a significant change in pattern on the flukes between sightings, or may simply be the result of a matching error. Both are more likely to occur if the fluke is originally less distinct (Friday et al. 2000). One study of Atlantic humpback whales demonstrated that over eight years, 4.6 % of the 152 flukes analyzed showed a major pattern change, 31.8 % showed a moderate change, and 63.6 % showed little or no change (Carlson et al. 1990). Given the particularly long temporal scale of the sightings used in this analysis, it is likely that some portion of these flukes underwent a degree of change in fluke pattern, contributing to the misidentification error. Misidentification is equivalent to a tag loss in a mark-recapture setting, which works to inflate population size estimates. If the degree of bias toward more distinct flukes was consistent over the time series, our ROI estimates would not be affected.

Studies that have specifically addressed bias from humpback whale photo identification have shown that misidentification and heterogeneity error is small, and can be mitigated for. A double tagging study by Stevick et al. (2001) showed that false positive errors were absent altogether and false negative errors were negligible (0.1 %). Heterogeneity was assessed in a study by Friday et al. (2008) where the authors concluded that the most accurate results could be found by simply removing poor quality photos. In the present study, we implemented a photo quality standard and built models including fluke distinctiveness covariates and found that they were not selected based on AICc values. Consequently, we conclude that the overall influence from fluke distinctiveness and heterogeneity in this study was minimal.

## 4.3.2 Jolly-Seber models

In our Jolly-Seber mark-recapture model, an overall increasing trend in abundance demonstrates that the western PWS humpback population is in recovery from the past over-exploitations of whaling with a 5.0-fold increase in abundance. However, there appears to be a decline from approximately 2000 to 2006. The cause of this decline is unknown, though it could represent a temporary emigration from the study area. Other studies have noted that while humpback whales demonstrate a general fidelity to the feeding grounds, it is common for them to move between feeding areas with the magnitude and frequency of these movements varying from individual to individual (Baker et al. 1992; Calambokidis et al. 1996, 2001, 2008; Straley et al.2008; Waite et al. 1999). Further, humpback whales are known to be generalist feeders (Nemoto and Kasuya1965; Witteveen et al. 2008), and it is commonly believed that their movements within feeding grounds can be linked to food availability. In a Gulf of Maine study (Weinrich et al. 1997), movements in Atlantic humpback whales were directly linked to food availability. Humpback whale abundance was documented to decline by over 95 % in one area and increase dramatically in a neighboring area, simultaneously to the same geographic shift in a humpback whale prey species.

The abundance estimate for 2009 is much higher than the overall trend in the model, but at this point it is unclear if this is a reflection of variability in the data or a surge in immigration into the area. There was an exceptionally high number of unique individuals identified in 2009 (n=89n=89), which leads us to believe that this high abundance estimate is, at least in part, driven by an increased number of individuals coming into the study area (immigration). Sightings of new individuals were seen in each year.6 In a closed population, sightings of new individuals would decrease over time as all individuals are discovered. However, in this open population, we see that the discovery of new individuals persists through the time series. Particularly, in this last year, 2009, we see an exceptionally high number of new sightings.

The overall abundance, estimated by the MARK software (P) was 396 (95 % CI 371–431). The number of unique individuals that went into the analysis was 322, indicating that the model estimates approximately 74 whales that were in the area at some point in the study, but were undetected.

### 4.3.3 Sensitivity analysis: early year photo quality

Because of the high correlation found in this sensitivity analysis, we concluded that the influence of poor quality photos in some early years is negligible in their effect on later years in the full time series models. However, it is possible that photos from these years are altering the results of the years they are in (1980, 1983, 1984, 1985, and 1987). In these years, inclusion of poor quality photos may have increased the risk of false negative errors-instances where photos are wrongly designated as new sightings—which may have led to an inflation of the population estimates. Though, without a way to estimate any direct or proxy measure of false negative errors in these years, we cannot measure the magnitude of this effect. Inflated estimates in these years would lead to an overall underestimation of ROI in this study. This means that our estimate of ROI could be biased low for this reason.

## 4.3.4 Sensitivity analysis: survival

In the selected model, the estimate of ϕϕ (set constant throughout the time series) was 0.94. This is lower than the only other humpback whale survival rate estimate from PWS made by Mizroch et al. (2004) of 0.98 (95 % CI 0.95–1.0). However, it is consistent with other estimates made by the same research group, namely: 0.93 (95 % CI 0.91–0.95) for all of Alaska, 0.94 (95 % CI 0.92–0.95) for Southeast Alaska, and 0.96 (95 % CI 0.94–0.98) for Hawaii (Mizroch et al. 2004). Further, our estimates are consistent with a survival estimate of 0.95 (95 % CI 0.93–0.97) made for North Atlantic humpback whales (Buckland 1990), and within range of the 0.96 (SE = 0.008) survival estimate also made for North Atlantic humpback whales (Barlow and Clapham 1997). While humpback whales exhibit some level of feeding site fidelity, many studies have documented movements that contradict complete fidelity (Baker et al. 1992; Calambokidis et al. 1996, 2001,2008; Straley et al. 2008; Waite et al. 1999). This is true for our study area as well; other analyses from our same dataset show clear evidence of exchange of individuals with adjacent areas (von Ziegesar and Matkin1986; von Ziegesar 2012). This means that because we have no way of distinguishing between death and emigration in our models, our estimates of survival may be incorporating emigration caused by movements in addition to actual mortalities. That said, we believe our estimate to be a good representation of the averaged survival over the entire time series for western PWS.

## 4.3.5 Comparison to previous abundance estimates

Humpback whales in PWS were estimated using a Chapman estimator at 100 (SE = 20) individuals in 1984 (Matkin and von Ziegesar 1985). Our Jolly-Seber abundance estimate from the same year was slightly lower, at 73 (95 % CI 54–99), although the confidence intervals for these estimates nearly overlap. The difference in these estimates is most likely due to the fact that photo quality was not considered in the previous study, and the inclusion of poor quality photos leads to inflated population estimates.

We estimated humpback whale abundance at 134 (95 % CI 114–158) and 194 (95 % CI 162–231), for the summer of 2008 and 2009, respectively, while an independent study estimated 135 (SE = 12) humpback whales in the 2008–2009 winter season using a Huggins closed capture model (Rice et al. 2011). It is important to note that these abundance estimates are not directly comparable because our data are from the summer and only from western PWS, while their data were collected in the winter and represent sound-wide observations. We would expect abundance to drop substantially in winter months as the vast majority of humpback whales are thought to spend winter months in the tropical breeding grounds (Darling and McSweeney 1985; Johnson and Wolman 1984). So, despite the difference in spatial coverage, it is somewhat surprising that estimates from both studies are of similar magnitude given the seasonal differences. This suggests that a large portion of summer whales are staying late into the winter season, and is supported anecdotally by sightings of individuals seen in both summer and into fall and winter (Moran 2011, pers. comm.). One theory is that these late-season humpback whales are staying in PWS to capitalize on the local winter stocks of Pacific herring (Rice et al. 2011).

## 4.3.6 Model performance

In our analysis of this humpback whale dataset of exceptionally long time-frame, various limitations in model estimation were discovered. First, we were unsuccessful in estimating time-dependent survival, ϕtϕt, in our mark-recapture models due to a high portion of the parameters converging on the boundary of one rather than producing realistic estimates. This is likely due to the particularly long life span of humpback whales (Gabriele et al. 2010). Because humpback whales are living so much longer than the time period between sampling intervals, the model is not able to generate fine-scale estimates of φφ during these times.

After managing the specific problematic parameters, these long-term data were successfully described by realistic parameter estimates. Therefore, we feel that the Jolly-Seber model is an appropriate option for analyzing long-term humpback whale datasets. With this study, the first time series of humpback whale abundance from 1978–2009 has been constructed and should be of use for further investigations of ecological processes in PWS.

## 4.4 Rate of increase

As previously mentioned, the SPLASH study generated estimates of abundance for humpback whales in the North Pacific Basin (Calambokidis et al. 2008). These estimates are biologically feasible given they are below the maximum theoretical ROI estimate of 11.8 % (Zerbini et al. 2010). Given the earlier estimates of humpback whale abundance in the North Pacific, the population has been recovering at a rate of 5–7 % annually in order to reach current North Pacific population estimates. In PWS, we found a slightly lower ROI, 4.53 % (95 % CI 3.28–5.79 %) where the 95 % CI of overlaps with the lower end of the SPLASH ROI estimate (4.9 %). The lower end of the SPLASH ROI estimate was found using more recent data, 1993–2006; because these years are overlapping with our time series, it makes sense that our estimate would align closer with the lower end of the SPLASH estimate. The upper end (6.8 %), on the other hand, uses data spanning back to 1966—over a decade before the start of our time series. Further, estimates of ROI from SPLASH are generally higher for the western Pacific regions, and between 5.5 and 6.0 % when assessing the population without considering the western Pacific (Calambokidis et al. 2008). Therefore, the ROI results are consistent with the broader trends of humpback whale recovery seen for the North Pacific. Thus, like the North Pacific population, the western PWS population has been rapidly recovering from past over-exploitation and may be reaching levels at or even exceeding those before commercial exploitation.

## FOOTNOTES

1See Electronic Supplementary Materials, Figure 1, for descriptions of these criteria.

2See Electronic Supplementary Materials, Figure 2, for descriptions of these criteria.

3See Electronic Supplementary Materials, Figure 3, for a graph of these data.

4See Electronic Supplementary Materials, Figure 4, for a graph of these results.

5See Electronic Supplementary Materials, Figure 5, for a graphic comparison between Jolly-Seber estimates and sightings per day.

6See Electronic Supplementary Materials, Figure 6, for a discovery curve that shows more detail on number of new sightings by year.

## ACKNOWLEDGMENTS

We thank Beth Goodwin, Shelley Gill, and Gerry Sanger for their substantial data contributions and to Woods Hole Oceanographic Institute for the contribution of the 1978–1979 data to the database used in this analysis. We are grateful to John Moran, Ron Heintz, and Jeep Rice at NMFS, Auke Bay Labs for their collaboration and to Nicola Hillgruber at UAF for advice and insight and to the reviewers of this manuscript who provided very helpful comments. Also, we acknowledge funding support from the Exxon Valdez Oil Spill Trustee Council (Project 10100128) and the Cooperative Institute for Arctic Research with funds from the National Oceanic and Atmospheric Administration under cooperative agreement NA17RJ1224 with the University of Alaska.

## REFERENCES


1. Baker SC, Straley JM, Perry A (1992) Population characteristics of individually identified humpback whales in southeastern Alaska: summer and fall 1986. Fish Bull 90:429–437

2. Barlow J, Clapham PJ (1997) A new birth-interval approach to estimating demographic parameters of humpback whales. Ecology 78:535–546

3. Buckland ST (1990) Estimation of survival rates from sightings of individually identifiable whales. Rep Int Whal Comm Special Issue 12:149–153

4. Burnham KP, Anderson DR (2002) Model selection and multi-model inference: a practical information-theoretical approach, 2nd edn. Springer, Berlin

5. Calambokidis J, Steiger GH, Evenson JR, Flynn KR, Balcomb KC, Claridge DE, Bloedel P, Straley JM, Baker CS, von Ziegesar O, Dahlheim ME, Waite JM, Darling JD, Ellis G, Green GA (1996) Interchange and isolation of humpback whales off California and other north pacific feeding grounds. Marine Mamm Sci 12(2):215–226

6. Calambokidis, J, Steiger GH, Straley JM, Quinn TJ, Herman LM, Cerchio S, Salden DR, Yamaguchi M, Sato F, Urban JR, Jacobsen J, von Zeigesar O, Balcomb KC, Gabriele CM, Dahlheim ME, Higashi N, Uchida S, Ford JKB, Miyamura Y, Ladron de Guevara P, Mizroch SA, Schlender L, Rasmussen K (1997) Abundance and population structure of humpback whales in the North Pacific basin. Final Contract Report 50ABNF500113 to Southwest Fisheries Science Center, P.O. Box 271, La Jolla, CA 92038, 72 pp

7. Calambokidis J, Steiger GH, Straley JM, Herman LM, Cerchio S, Salden DR, Urban RJ, Jacobsen JK, von Ziegesar O, Balcomb KC, Gabriele CM, Dahlheim ME, Uchida S, Ellis G, Mlyamura Y, Ladron

PP, Manami Y, Sato F, Mizroch SA, Schlender L, Rasmussen K, Barlow B, Quinn TJ (2001) Movements and population structure of humpback whales in the North Pacific. Marine Mamm Sci 17(4):769–794

8. Calambokidis J, Falcone EA, Quinn TJ, Burdin AM, Clapham PJ, Ford JKB, Gabriele CM, LeDuc R, Mattila D, Rojas-Bracho L, Straley JM, Taylor BL, Urban J, Weller D, Witteveen BH, Yamaguchi M, Bendlin A, Camacho D, Flynn K, Havron A, Huggins J, Maloney N (2008) SPLASH: structure of populations, levels of abundance and status of humpback whales in the North Pacific. Final report for contract AB133F-03-RP-00078, U.S. Dept of Commerce Western Administrative Center, Seattle, WA, 57 pp

9. Carlson CA, Mayo CA, Whitehead H (1990) Changes in the ventral fluke pattern of the humpback whale*(Megaptera novaeangliae)*, and its effect on matching; evaluation of its significance to photo-identification research. Rep Int Whal Comm Special Issue 12:105–111

10. Darling JD, McSweeney DJ (1985) Observations on the migrations of North Pacific humpback whales*(Megaptera novaeangliae)*. Can J Zool 63(2):308–314

11. Friday NA, Smith TD, Stevick PT (2000) Measurement of photographic quality and individual distinctiveness for the photographic identification of humpback whales, *Megaptera novaeangliae*. Marine Mamm Sci 16(2):355–374

12. Friday NA, Smith TD, Stevick PT, Allen J, Fernald T (2008) Balancing bias and precision in capture-recapture estimates of abundance. Marine Mamm Sci 24(2):253–275

13. Gabriele CM, Lockyer C, Straley JM, Jurasz CM, Hidehiro K (2010) Sighting history of a naturally marked humpback whale *(Megaptera novaeangliae)* suggests ear plug growth layer groups are deposited annually. Marine Mamm Sci 26(2):443–450

14. Gambell R (1976) World whale stocks. Mamm Rev 6:41–53

15. Johnson JH, Wolman AA (1984) The humpback whale (*Megaptera novaeangliae*). Marine Fish Rev 46(4):30–37

16. Katona S, Baxter B, Brazier O, Kraus SD, Perkins J, Whitehead H (1979) Identification of Humpback Whales by fluke photographs. In: Winn HE et al (1979) Behavior of marine animals: current perspectives in research, vol 3. Cetaceans. Plenum Press, New York, pp 33–44

17. Lebreton JD, Burnham KP, Clobert J, Anderson DR (1992) Modeling survival and testing biological hypotheses using marked animals: a unified approach with case studies. Ecol Monogr 62:67–118

18. Matkin CO, von Ziegesar O (1985) Humpback whales in Prince William Sound in, (1984) Report for contract 41 USC 252. National Marine Fisheries Service, National Marine Mammal Laboratory, Seattle 14 pp

19. Mizroch SA, Herman LM, Straley JM, Glockner-Ferrari DA, Jurasz C, Darling J, Cerchio S, Gabriele CM, Salden DR, von Ziegesar O (2004) Estimating the adult survival rate of central North Pacific humpback whales*(Megaptera novaeangliae)*. J Mammal 85(5): 963–972

20. Nemoto T (1970) Feeding patterns of baleen whales in the oceans. In: Steele JH (ed) Marine Food Chains. University of California Press, Berkeley, pp 241–252

21. Nemoto T, Kasuya T (1965) Foods of baleen whales in the Gulf of Alaska of the North Pacific. Sci Rep Whales Res Inst 19:45–51

22. Pradel R (1996) Utilization of capture-mark-recapture for the study of recruitment and population growth rate. Biometrics 52:703–709

23. Rice DW (1978) The humpback whale in the North Pacific: distribution, exploitation, and numbers. In: Norris KS, Reeves R (eds) Report on a workshop on problems related to humpback whales *(Megaptera novaeangliae)* in Hawaii. Report to the Marine Mammal Commission, July 1977, Washington, DC, 21 pp

24. Rice SD, Moran JR, Straley JM, Boswell KM, Heintz RA (2011) Significance of whale predation on natural mortality rate of Pacific herring in Prince William Sound, Exxon Valdez Oil Spill Restoration Project Final Report (Restoration Project: 100804). National Marine Fisheries Service, Juneau

25. Schwarz CJ (2001) Jolly-Seber model: more than just abundance. J Agric Biol Environ Stat 6:195–205

26. Schwarz CJ, Arnason AN (1996) A general methodology for the analysis of capture-recapture experiments in open populations. Biometrics 52:860–873

27. Seber GAF (1982) The estimation of animal abundance, 2nd edn. Wiley, New York

28. Stevick PT, Palsoll PJ, Smith TD, Bravington MV, Hammond PS (2001) Errors in identification using natural markings: rates, sources, and effects on capture-recapture estimates of abundance. Can J Fish Aquatic Sci 58:1861–1870

29. Straley JM (1990) Fall and winter occurrences of humpback whales *(Megaptera novaeangliae)* in southeastern Alaska. Rep Int Whal Comm Special Issue 12:SC/A88/ID26

30. Straley JM (1994) Seasonal characteristics of humpback whales *(Megaptera novaeangliae)* in southeastern Alaska. Thesis, University of Alaska, Fairbanks

31. Straley JM, Quinn TJ II, Gabriele CM (2008) Assessment of mark-recapture models to estimate the abundance of a humpback

whale feeding aggregation in Southeast Alaska. J Biogeogr Special Issue 36:427–438

32. von Ziegesar O (1992) A catalog of Prince William Sound humpback whales by fluke photographs between the years 1977 and 1991. Prepared by North Gulf Oceanic Society for the National Marine Fisheries Service, National Marine Mammal Laboratory Seattle, WA

33. von Ziegesar O (2010) A catalog of humpback whales In Prince William Sound and Kenai Fjords, Alaska 1977–2010. Printed and distributed by Eye of the Whale Research, P.O. Box 15191, Fritz Creek, AK 99603

34. von Ziegesar O (2012) The Humpback whales of Prince William Sound: population characteristics and ecology, 1980–2009. Report to the Exxon Valdez Trustees Council

35. von Ziegesar O, Matkin C (1985) A catalog of humpback whale flukes in Prince William Sound, Alaska identified by fluke photographs between the years 1977 and 1984. Report on contract no. 41 USC 252. National Marine Fisheries Service, National Marine Mammal Laboratory Seattle, WA

36. von Ziegesar O, Matkin C (1986) Humpback whales in Prince William Sound in 1985. Report on contract no. 41 USC 252. National Marine Fisheries Service, National Marine Mammal Laboratory Seattle, WA

37. von Ziegesar O, Goodwin B, Devito R (2001) A catalog of humpback whales In Prince William Sound, Alaska 1977–2001. Printed and distributed by Eye of the Whale Research, P.O. Box 15191, Homer, AK 99603

38. Waite JM, Dahlheim ME, Hobbs RC, Mizroch SA, von Ziegesar-Matkin O, Herman LM, Jacobsen J (1999) Evidence of feeding

aggregation of humpback whales *(Megaptera novaeangliae)* around Kodiak Island, Alaska. Marine Mamm Sci 15:210–220

39. Weinrich M, Martin M, Griffiths R, Bove J, Schilling M (1997) A shift in distribution of humpback whales,*(Megaptera novaeangliae)*, in response to prey in the southern Gulf of Maine. Fish Bull 95(4):826–836

40. White GC, Burnham KP (1999) Program MARK: survival estimation from populations of marked animals. Bird Study 46(Supplement):120–138

41. Witteveen BH, Foy RJ, Wynne KM (2006) The effect of predation (current and historical) by humpback whales*(Megaptera novaeangliae)* on fish abundance near Kodiak Island, Alaska. Fish Bull 104:10–20

42. Witteveen BH, Foy RJ, Wynne KM, Tremblay Y (2008) Investigation of foraging habits and prey selection by humpback whales *(Megaptera novaeangliae)* using acoustic tags and concurrent fish surveys. Marine Mamm Sci 12(3):516–534

43. Zerbini AN, Clapham PJ, Wade PR (2010) Assessing plausible rates of population growth in humpback whales from life-history data. Marine Biol 157(6):1225–1236

8

# MANAGING PLANTED FORESTS FOR MULTIPLE USES UNDER A CHANGING ENVIRONMENT IN CHINA

**Shirong Liu[1*], Shuirong Wu[2], Hui Wang[1]**

[1]Institute of Forest Ecology, Environment and Protection, Chinese Academy of Forestry, Beijing 100091, PR. China.

[2]Research Institute of Forestry Policy and Information, Chinese Academy of Forestry, Beijing 100091, PR. China.

## ABSTRACT

### Background

Planted forests are expanding throughout the world, and now account for 7% of global forest cover and provide more than 60% of global industrial round wood. Negative ecological and social impacts of the establishment of planted forests and the challenges of their multi-purpose management have also given rise to concern. China has been playing an important role in global expansion of planted forests while reducing emission from deforestation and forest degradation.

### Methods

This article attempts to conduct an overall analysis and review of the current status, challenges and future perspectives of planted forests in China to obtain a better understanding on how to manage planted forests for multiple uses under a changing environment. Data from several national forest inventories and other sources, as well as new empirical data, were used for a statistical analysis on the dynamics of planted forests in China.

## Results

Planted forests in China have undergone a continuous expansion in the past 20 years, which has significantly contributed to an increase in total forest cover and timber supply as well as other ecosystem services like carbon sequestration. The three key driving forces for this expansion were government programmes, and market and technology development. However, the predominance of very few tree species in the plantations, uneven spatial distribution, skewed age-class distribution, and low volumes in growing stock, coupled with increasing complexity of multiple purpose forestry management under a changing environment, have generated several major challenges confronting planted forests in China.

## Conclusions

A strategic transition in the management of Chinese planted forests is needed, with a shifting emphasis from area expansion to stand productivity and quality enhancement, from traditional timber production to multi-purpose management for forest goods and services, and from monoculture plantations to biodiversity rich mixed forests. A landscape-design approach and adaptive management practices should be put in place to meet the diversified demands of stakeholders for different goods and ecosystem services while enhancing forest resilience under the changing climate.

# 1. BACKGROUND

Planted forests are composed of trees established through planting and/or through deliberate seeding of native or introduced species for productive and protective purposes. Such forests have been expanding rapidly in all regions of the world reaching 264 million ha in 2010 (FAO, 2010), of which nearly 48% was established within the past 20 years. These planted forests are mainly distributed in Asia, accounting for 46% of the total world planted forest area, followed by 26% in Europe and 14% in North America. China alone accounts for 23% of the total planted forest area in the world, with a rapid expansion rate of 1.76 million ha annually, contributing significantly to the offsetting of annual net global loss of 6.77 million ha in the past 20-year period (FAO, 2010). Besides the expansion of the planted forests, the natural forests in China have been also increasing with a net gain of 9.57 million ha since 1998 when China launched the Natural Forest Protection Program (State Forestry Administration (referred as SFA below), 2014).

Although planted forests represent only 7% of the global forested area, their contribution to the global industrial round wood production exceeds 60% (Carle and Holmgren, 2008). It is expected that the expansion of planted forests will continue or even accelerate in many parts of the world in order to meet the increasingly diversified demands on forest goods and services from the society. Given the current expansion trend, a further rise in the area of planted forests up to 300 million hectares by 2020 can be anticipated (FAO, 2010). Planted forests are therefore expected to increasingly contribute to the world's supply of wood, fibre, fuel and non-timber forest products (NTFPs), as well as provide environmental and social services. In particular, the central role of planted forests in mitigating climate changes, conserving biodiversity, combating land degradation and developing green economy has been increasingly recognised in the evolving international processes and financing mechanisms (Winjum and Schroeder, 1997; Paul et al, 2012). At the same time, negative ecological and social impacts of planted forests and the challenges confronting planted forests have also given rise to serious concern across the globe (Winjum and Schroeder, 1997; Booth, 2013; Ferraz et al, 2012).

Due to its important role in global expansion of planted forests, China's experience provides a deep insight into the mechanisms underlying the successful development of planted forests. The impacts of planted-forest development on timber markets and the environment are of significance to policy-makers, planners and forest managers. A number of studies carried out in the world predict the future contribution of planted forests to economic, environmental and social services. However, there have been no comprehensive analyses on the development of planted forests in China so far, even though global assessments of planted forest resources and outlook studies for wood from planted forests, best practice guidelines and many interesting cases of country applications from around the world have been documented (FAO, 2006a, 2006b; Kröger, 2012; Carle and Holmgren, 2008; Cossalter and Barr, 2005; FAO, 2010). This paper presents an overview on the current status and driving forces of planted forest development in China, and describes the major challenges and future perspectives of China's planted forests. The authors' aim is that this article will provide useful information and experiences for other developing countries, and assist with understanding the overall situation of the planted forest resource in China and its future development strategies under climate change.

## 2. METHODOLOGY

The method applied in this paper is basically analysis of the available data and other information. The data used are publicly available except where specifically stated. Most of data used for the analysis originate from the Food and Agriculture Organization (FAO) statistics, the national forest inventories and the forestry statistics by the Chinese Government referenced herein. The FAO statistics, mainly the Global Forest Resources Assessment 2010, provide a good database for global and regional analysis. Forest resources inventories have been conducted periodically in China for about 40 years. The data from eight times national forest inventories have been used to explore the dynamics of planted forests in China using information on tree species, age structure and spatial distribution and so on. The data on planting area, investments and so on have been derived from the Chinese Forestry Statistical Yearbook published annually by the Government. In order to explore the forces driving establishment of new forests at local level, an economic investigation was conducted in 2011 for a comparative cost-benefit analysis through interviews with 50 farmers from the Banqiao village and the Guantang village in the city of Qinzhou in the Guangxi region of China where*Eucalyptus* spp. plantations have been greatly expanded in the last decade.

## 3. CURRENT STATUS OF PLANTED FORESTS IN CHINA

Historically, China was rich in forest resources and biodiversity due to its vast geographical areas and highly diversified climate conditions that provided favourable environments for widely varying forest ecosystems ranging from boreal forests in the north to tropical rain forests in the south. Most forests were public assets and strong centralised governance ensured their strict protection against unauthorised occupation and illegal logging except during the declining period of Qing dynasty in the latter part of nineteenth century and the first half of the twentieth century. Post revolution, China's forestry development has experienced an oscillating process, as in most places across the world (Lane and McDonald, 2002), characterised by three distinct phases of: (i) over-logging and deforestation resulting in reduction in the extent and quality of forest cover, (ii) massive reforestation and afforestation across the country with a few species for increased production of timber and other forest products, and (iii) shifting focus of management from production-centred forestry towards higher emphasis on ecological rehabilitation, environmental amelioration and provision of ecosystem services (S. Liu et al., 2011).

During the latter two phases, afforestation and reforestation have played a dominant role in promoting the rapid and massive expansion of planted forests in China (Figure 1). The most recent national forest inventory places China's forest cover at 21.63% of the country's geographical area, up from just about 10% at the end of 1970s. China today has about 208 million ha of forested land and 15.14 billion m³ of stocking volume, of which the total planted forests account for 69.33 million ha (36%) of the total forest area and 2.48 billion m³ (17% ) of the total forest stocking volume of the country (SFA, 2014).

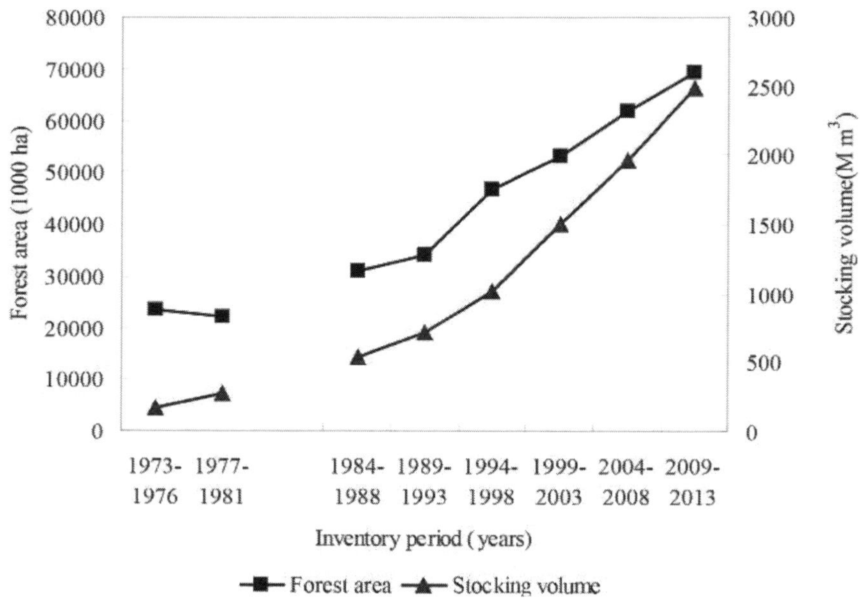


**Figure 1. Area and volume of planted forests in China over the last forty years**. Sources: Data from the eight national forest inventories in China, including General Bureau of State Forestry, 1977; Ministry of Forestry, 1983, 1989, 1994; SFA, 2000,2005, 2009, 2014.

## 3.1 Types of planted forests

China's Forest Law (1998) classifies the five major types of forests based on their management objectives, namely: (i) timber forest, with the main purpose of producing timber, including bamboo forests which are utilised for bamboo wood; (ii) economic forest, with the main management emphasis on forest products other than timber, such as fruit, nuts, herbs, bark, leaves, tree sap, flower buds and tender sprouts, and raw materials for forestry chemical industries etc. (In China's forest inventory system, economic forest is measured only in terms of area, but not stocking volume.); (iii) fuel-wood forest, with the main purpose of producing fuel wood; (iv) protection forest, with the main emphasis on protection and provision of services; and (v) special use forest, with the purpose of special use such as national defence, environmental conservation, scientific experiments etc.

The first three types are categorised as commercial forests, which are predominantly managed for economic purposes, and the latter two are categorised as ecological forests predominantly managed for ecological purposes. Among the five types, timber forest accounts for 43% of the total planted forest area, followed by economic forest and protection forest. In terms of stocking volume, timber forest accounts for 63% of total planted forests, followed by protection forest (Table1).

**Table 1.** Composition of types of planted forests in China

| Forest type | Area | | Stocking volume | |
|---|---|---|---|---|
| | 1000 ha | % | 1000 $m^3$ | % |
| Timber | 29,753 | 42.9 | 1,551,938 | 62.5 |
| Protection | 17,993 | 26.0 | 840,375 | 33.8 |
| Special use | 1,473 | 2.1 | 87,288 | 3.5 |
| Economic | 19,855 | 28.6 | - | 0.0 |
| Fuelwood | 260 | 0.4 | 3,647 | 0.1 |
| Total | 69,334 | 100 | 2,483,248 | 100 |

*Sources: The Eighth China's National Forest Inventory* (SFA, 2014).

## 3.2 Dominant tree species in planted forests

According to the characteristics of trees in terms of height and trunk, the planted forests are classified as high forests, economic forests and bamboo forests in China's forest inventory (Xiao, 2005). High forest is the one that originates from seed or from planted seedlings, with standalone trunks and height over six metres. Economic forest as one of the five major types of forests is separated from the high forest when describing the dominant tree species. Bamboo forest refers to the forest land spanning more than 0.067 ha (one Chinese moo) and growing bamboo species with a diameter at breast height of over 2 cm. The latest national forest inventory shows that the planted forests consist of 47 million ha of high forests, 19.8 million ha of economic forests and 2.4 million ha of bamboo forests.

Ten tree species dominate the 34.3 million ha of planted high forests and they contain 1.87 billion $m^3$ of stocking volume according to the latest National Forest Inventory, Table 2. The remaining 12.7 million ha consists of more than 60 tree species. Among these dominant species, the top five are Chinese fir (*Cunninghamia lanceolata* (Lambert). Hooker), poplars (*Populus* spp.), eucalyptus (*Eucalyptus robusta* Smith), larch (*Larix gmelinii* Rupr.) and Masson's pine (*Pinus massoniana* Lamb.), accounting for 60% of total area of planted high forest and 66% of total volume of planted high forest (Table 2) respectively. Following attempts to broaden the ecological base of plantations in recent years, mixed plantations now cover 16% of planted forests while the rest is monoculture.

## 3.3 Age classes of planted forests

In the forest inventory of China, there are five age classes that are young, middle- aged, pre-mature, mature, and over-mature. The classification of age class follows national guidelines for forest resources survey (SFA, 2011a). The age class interval can be divided by 20 years, 10 years, 5 years, or 1~2 years depending on the growth rate and life time of various tree species. Since major reforestation and afforestation work in China began in the 1970s and gathered momentum thereafter, the young and middle-aged classes dominate the forest landscape accounting for nearly 72% of the total area of planted high forest and 52% of the total stocking volume of planted high forest (Table 3), respectively.

**Table 2.** Composition of dominant tree species (group) in Chinese planted high forests

| Dominant tree species | Area | | Stocking volume | |
|---|---|---|---|---|
| | 1000 ha | % | 1000 $m^3$ | % |
| *Cunninghamia lanceolata* | 8,946 | 19.0 | 625,404 | 25.0 |
| *Populus* spp. | 8,538 | 18.1 | 502,958 | 20.1 |
| *Eucalyptus robusta* | 4,455 | 9.5 | 160,334 | 6.4 |
| *Larix gmelinii* | 3,137 | 6.7 | 184,142 | 7.4 |
| *Pinus massoniana* | 3,062 | 6.5 | 171,552 | 6.8 |
| *Pinus tabulaeformis* Carr. | 1,608 | 3.4 | 66,072 | 2.6 |
| *Cupressus funebris* | 1,462 | 3.1 | 61,042 | 2.4 |
| *Pinus elliottii* Engelm. | 1,344 | 2.9 | 40,560 | 1.6 |
| *Robinia pseudoacacia* L. | 1,226 | 2.6 | 26,992 | 1.1 |
| *Quercus* spp. | 541 | 1.1 | 34,396 | 1.4 |
| Others | 12,745 | 27.1 | 631,735 | 25.2 |
| Total | 47,064 | 100 | 2,505,187 | 100 |

*Sources: The Eighth China National Forest Inventory* (SFA, 2014).

## 3.4 Spatial distribution of planted forests

The planted forests in China are unevenly distributed across the whole country, with greater concentration in the warmer and wetter south and southwest regions, accounting for 63% of the total area and 62% of the total stocking volume of planted forests, respectively (Table 4). The top five provinces are Guangxi, Guangdong, Hunan, Sichuan and Yunnan, reaching 25.3 million ha, accounting for 37% of the total area and 31% of the total stocking volume of planted forests, respectively.

## 3.5 Ecosystem goods and services from planted forests

Planted forests, like natural forests and many other types of vegetation, provide ecosystem goods and services that significantly contribute to human well-being. Currently, the annual harvest from planted forests reaches 155 million cubic metres, accounting for 46% of the total wood production (SFA, 2014). Being one of the leading timber consumers in the world, China's wood demand has been rapidly growing with the increasing national economy and population. The increasing gap between China's timber supply and demand has to be mostly fulfilled from its own planted forests although some timber is imported from planted forests in countries such as New Zealand.

**Table 3.** Age structure of planted high forests

| Age class | Area | | Stocking volume | |
|---|---|---|---|---|
| | 1000 ha | % | 1000 $m^3$ | % |
| Young | 18,662 | 39.6 | 357,099 | 14.4 |
| Middle-aged | 15,147 | 32.2 | 927,257 | 37.3 |
| Close to mature | 6,679 | 14.2 | 581,627 | 23.4 |
| Mature | 5,099 | 10.8 | 483,939 | 19.5 |
| Post-mature | 1,483 | 3.2 | 133,327 | 5.4 |
| Total | 47,070 | 100.0 | 2,483,249 | 100.0 |

*Sources: The Eighth China National Forest Inventory* (SFA, 2014).

Besides timber products, planted forests also contribute to economic development and livelihood by providing NTFPs. Of the total economic forests of the country, 96% is planted forests, producing 133.8 million tons of NTFPs in 2011 (of which, 114.7 million tons of fruits, 9.3 million tons of dry fruits/nuts, 1.4 million tons of tea, 1.6 million tons of woody oil and etc), accounting for 23% of the total output value of forestry industry (SFA, 2011b).

At the same time, planted forests play a key role in ecological improvement by enhancing environmental services of biodiversity conservation, carbon sequestration and hydrological regulation greatly contributing to the China's strategic goals of ecological restoration, ecological security and ecological civilisation (SFA, 2009). The monetary value of the major forest ecosystem services such as water conservation, soil conservation, carbon sequestration and oxygen release, nutrient accumulation, atmosphere environment purification, and biodiversity conservation, was estimated to be about 10.01 trillion CNY/year (approximately 1.48 trillion US$/year) and is equivalent to one third of China's annual gross domestic product (GDP) (Niu et al, 2012). Of these, carbon (C) sequestration by planted forest is the most important service with about 0.3 Pg C having been sequestrated by planted forests in China since the mid-1970s. The rapidly increasing C stock (and C density) in China's planted forests has been a major contribution to the increase in the global forest C sink over the period (Fang et al., 2001; Piao et al., 2009; Pan et al., 2011). Since most of these planted forests are still very young and have not always benefited from the greatly increased public investment in forest technology and management in China, it is reasonable to assume that the full potential of planted forests in China to accumulate additional biomass in living trees and in forest soils has not yet been achieved (Fang et al., 2007).

**Table 4.** Spatial distribution of planted forests in China

| Region and province | Area | | Stocking volume | |
|---|---|---|---|---|
| | 1000 ha | % | 1000 $m^3$ | % |
| **South and southwest region** | **43,962** | **63.3** | **1,546,271** | **62.3** |
| Guangxi | 6,345 | 9.2 | 222,722 | 9.0 |
| Guangdong | 5,579 | 8.0 | 154,677 | 6.2 |
| Hunan | 4,746 | 6.8 | 140,945 | 5.7 |
| Sichuan | 4,493 | 6.5 | 159,645 | 6.4 |
| Yunan | 4,141 | 6.0 | 110,203 | 4.4 |
| Fujian | 3,777 | 5.4 | 248,532 | 10.0 |
| Jiangxi | 3,386 | 4.9 | 111,219 | 4.5 |
| Zhejiang | 2,585 | 3.7 | 68,318 | 2.8 |
| Guizhou | 2,373 | 3.4 | 115,574 | 4.7 |
| Anhui | 2,251 | 3.2 | 93,748 | 3.8 |
| Hubei | 1,949 | 2.8 | 60,071 | 2.4 |
| Hainan | 1,362 | 2.0 | 23,132 | 0.9 |
| Chongqing | 926 | 1.3 | 35,917 | 1.4 |
| Tibet | 49 | 0.1 | 1,568 | 0.1 |
| **North and East Plain region** | **10,357** | **15.0** | **351,726** | **14.1** |

| | | | | |
|---|---|---|---|---|
| Shandong | 2,445 | 3.5 | 87,093 | 3.5 |
| Henan | 2,271 | 3.3 | 104,658 | 4.2 |
| Hebei | 2,209 | 3.2 | 56,838 | 2.3 |
| Jiangsu | 1,568 | 2.3 | 63,209 | 2.5 |
| Shanxi | 1,318 | 1.9 | 26,658 | 1.1 |
| Beijing | 372 | 0.5 | 7,857 | 0.3 |
| Tianjin | 106 | 0.2 | 3,549 | 0.1 |
| Shanghai | 68 | 0.1 | 1,864 | 0.1 |
| **Northeast region** | **10,459** | **15.1** | **461,068** | **18.5** |
| Inner Mongonia | 3,317 | 4.8 | 97,982 | 3.9 |
| Niaoning | 3,071 | 4.4 | 94,875 | 3.8 |
| Jilin | 2,465 | 3.6 | 164,237 | 6.6 |
| Heilongjiang | 1,606 | 2.3 | 103,974 | 4.2 |
| **Northwest region** | **4,558** | **6.6** | **124,187** | **4.9** |
| Shaanxi | 2,370 | 3.4 | 28,121 | 1.1 |
| Gansu | 1,030 | 1.5 | 28,318 | 1.1 |
| Xinjiang | 940 | 1.4 | 60,267 | 2.4 |
| Ningxia | 144 | 0.2 | 3,174 | 0.1 |
| Qinghai | 74 | 0.1 | 4,307 | 0.2 |
| **Total** | **69,336** | **100.0** | **2,483,252** | **100.0** |

*Sources: The Eighth China's National Forest Inventory* (SFA, 2014).

# 4. FACTORS SUSTAINING THE MASSIVE AFFORESTATION AND REFORESTATION IN CHINA

Since the 1970s, the planting area has been increasing rapidly with more than 5 million ha per year in average (Figure 2) without showing any signs of slowing even after almost four decades of continuous run, a rate so huge even for a country of the size of China that it could only have happened in an environment in which political, economic, ecological, social and aesthetic factors coalesced in resonance. In the beginning it was essentially government-initiated programmes that stimulated tree planting across the country in the face of severe depletion of forests after decades of mismanagement, severe misuse and complete neglect. Growth in urban and global markets for forest products, coupled with demographic migration from rural areas into urban zones, spurred the conversion of abandoned crop fields into tree farms. Quantitative, cross-national analyses suggest that many of these forces work jointly in regionally distinctive ways to facilitate the expansion of planted forest (Zhang et al, 2006; Rudel, 2009).

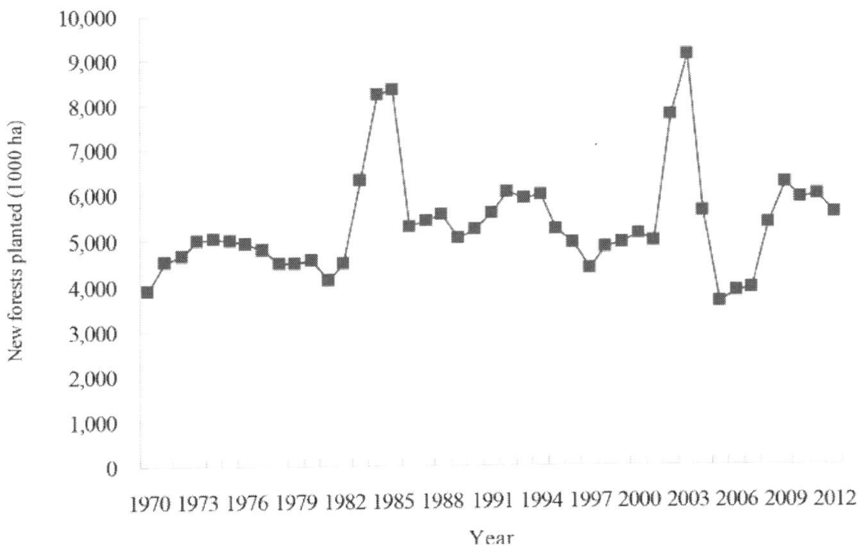


**Figure 2. Annual amount of new forests planted in China from 1970-2012**. *Sources: China Forestry Statistical Year Book, 2013a*

## (a) Government-initiated programmes

This is by far the most important driving force for stimulating the rapid and massive expansion of planted forests in China. Government initiated programmes either offer adequate financial supports or create an enabling environment for private investment in tree planting through attractive incentives. The implementation of the six key national forestry programmes in the past three decades, including the Key Shelterbelt Development Programs (SDP), the Natural Forest Protection Program (NFPP), the Conversion of Cropland to Forest Program (CCPF), the Sandification Control Program for Areas in the Vicinity of Beijing and Tianjin (SCP), the Wildlife Conservation and Nature Reserves Development Program, and the Forest Industrial Base Development Program (FIBDP), generated a great momentum for developing planted forests (S. Liu et al., 2011). Data for the period 1990 to 2012 showed that these six key forestry programmes have funded afforestation area of 78.2 million ha, contributing 62% of the total planted area (SFA, 2013a).

Over the last decade deep changes encompassing social, economic, ecological and political aspects have occurred in China generating an enabling environment for forest resource management at local, county, provincial and national levels. Among these perhaps the most important and far-reaching is forest tenure reform through which the management control right over 180 million ha of collective forestlands has already been passed into the hands of the individual households (SFA, 2013b). The Government hopes that this step will prove to be a great incentive and impetus to forest owners to further expand fast-growing and high-yielding planted forests for maximising economic benefit for themselves and, in the process, bring a wide range of ecological and social benefits to the larger society around them. While it is still too early to make informed judgments about the ecological outcome of one of the greatest steps ever undertaken in the ecological history of mankind, the interim indications are positive in nature, especially in south and southwest China where more favourable climatic conditions for tree growing occur. In addition to these land reforms, private investment in the forestry sector has also been increasing in all regions of China from 22% of the total forestry investment in 2004 when the reforms initiates at pilot scale to 63% of that in 2012 (SFA, 2013a), with private owners now managing more than 60% of the total planted forest area (SFA, 2014).

### (b) Market growth

Two key factors have stimulated tree planting campaign. These are the development of international and local markets for forest products and progressive urbanisation processes that bring improved transportation infrastructure and demographical immigration from rural areas to cities. In turn, this leads to the increased abandonment of farm lands due to less available labour. Also, Sohngen & Sedjo (1999) suggested that, under a varying price assumption, the increased demand for wood would be mostly met from planted forests due to the significantly lower cost of extraction. From a local perspective too, fast-growing planted forests have obvious cost-benefit advantages over cultivating annual agricultural crops. Data obtained from an economic investigation in Qinzhou city of Guangxi show that growing a single five-year-rotation of *Eucalyptus* spp. could yield 10,320 CNY/ha (1650 US$/ha) more net profit than cultivating annual crops over the same period (Table 5).

### (c) Technology development

Wider application of new technology is helping to improve the quantity and quality of trees, including improvement of growth rates, disease and pest resistance, tree form and wood-fibre quality, and ease of processing. It is also helping to expand the use of resources previously considered uneconomic (lesser known species, small piece sizes) and non-forest fibre resources such as wood residues and recycled materials (Wijewardana, 2005). Tissue culture and clonal reproduction are widely used in China as the most effective ways of providing a large quantity of high-quality seedlings for developing massive planted forests in relatively short periods (Lan and Gu, 2002). Increased yields can offer major economic advantages to planted forests based on biotechnology (Sedjo, 2004). Depending on the relative final returns of growing trees or wood processing, growing large trees over long rotations may be replaced with short-rotation management of small trees for production of fibre and biomass. Market competitiveness will be significantly influenced by the application of either traditional genetic and silvicultural-improvement practices, and wood processing techniques in planted forest development, or some new technical innovations beyond traditional forestry areas, such as new composite materials and bio-energy.

**Table 5.** A comparative cost-benefit analysis of annual crops (wheat and soybean) and *Eucalyptus* spp.

| a) Cost-benefit analysis of growing a wheat and soybean as an annual crop (Unit: CNY/ha) | | | |
|---|---|---|---|
| Two crops in a year: wheat + soybean | | | |
| **Cost** | **wheat** | | **Soybean** |
| Seed | 750 | | 750 |
| Fertiliser | 1800 | | 900 |
| pesticide/herbicide | 225 | | 225 |
| Plough | 600 | | 0 |
| Sowing | 525 | | 525 |
| Harvesting | 600 | | 600 |
| subtotal | 4500 | | 3000 |
| Total cost | | 7500 | |
| **Benefit** | **wheat** | | **Soybean** |
| yield per ha (kg) | 6000 | | 1500 |
| unit price | 2 | | 3.5 |
| subtotal | 12000 | | 5250 |
| Total benefit | | 17250 | |
| **Net profit per year per ha** | | **9750** | |
| **Net profit over 5 years** | | **48750** | |

| b) Cost-benefit analysis of growing Eucalyptus spp. (Unit: CNY/ha) | |
|---|---|
| **Costs of planting and management** | |
| Seedlings(0.5 CNY/plant×1500 plant/ha, including 150 plants for replanting) | 750 |
| Planting(0.6 CNY/hole×1500 holes/ha) | 900 |
| fertilisers (3.5 CNY/plant×1500 plants/ha) | 5250 |
| tending/fertilising(525 CNY/ha×4 times) | 2100 |
| subtotal | 9000 |
| **Costs of harvesting** | |
| harvesting and transportation(50 CNY/m$^3$ ×150 m$^3$/ha) | 7500 |
| Taxes and dues | 6930 |
| subtotal | 14430 |
| Total cost (over 5 years) | 23430 |
| **Benefits** | |
| Yield per ha in 5 years (m$^3$) | 150 |
| unit price | 550 |
| Total benefit (over 5 years) | 82500 |
| **Net profit** | |
| Total over 5 years | 59070 |

# 5. CHALLENGES CONFRONTING PLANTED FORESTS IN CHINA

Like many other countries in the world, planted forests in China have been exposed to huge challenges in terms of ecological sustainability and environmental changes. In particular, climate change has been regarded as a main disturbance driving force for the past several decades and future years to come.

## 5.1 Coping with uncertainties of climate change

Warming temperatures, changing precipitation patterns and extreme meteorological events are among possible threats to planted forests in China. Changes in temperature and precipitation regimes have the potential to gradually affect forests in terms of forest structure, spatial distribution, growth and productivity. Some effects from rising temperature and increasing precipitation, particularly in the colder north, may be positive for forest growth and productivity, while others (e.g. increased fire occurrence and pest and disease outbreaks) may be negative (S. Liu et al., 2011). In addition, changing climate can affect hydrological processes and water yields of forested watersheds, as well as the downstream water availability for both people and wetland ecosystems (Minshal, 1988; Poff and Ward, 1989; Poff, 1996; Sun et al., 2008). Climate extremes can be highly detrimental to forest ecosystems. For instance, the icy-rain and snowfall that occurred during the early spring of 2008 caused severe forest losses and destroyed nearly 13% of China's total forested area (Shao et al., 2011), which resulted in habitat loss and starvation of many kinds of wildlife besides potential, cumulative, and long-term effects on natural ecosystems. The recent sustained drought in Southwest China affected more than 80% of the vegetation in three provinces of Yunnan, Guangxi and Guizhou (Wang et al., 2010). If, as anticipated, the growing concern with global warming spurs further expansion in planted forests in an effort to sequester atmospheric carbon, questions about their social and ecological effects would become more pressing. The adaptive management of planted forests needs to consider the uncertainties of climatic change and its effects on forests and environments in order to enhance the positive effects while reducing the negative effects.

## 5.2 Preventing forest pest, disease and fire damage

The methods of establishing (by clear felling of natural forest before planting) and managing (using slash and burn) fast-growing monoculture plantations, predominantly of Masson pine, Chinese fir and *Eucalyptus robusta* in southern China and poplar species and larch in northern China, have caused a decline in biodiversity and consequential lowering of ecosystem stability and soil fertility. Many of these monoculture plantations are also in poor ecological health and are very vulnerable to exacerbating disturbances such as fire, insect and disease, hurricanes, drought, and wind storms under climate change (Peng et al., 2008). In particular, the disturbance agents of insect and pests, diseases and fire are expected to become the main challenges to forest managers. A total of 292 forest pests and diseases have been identified as of special concern to planted forests in China (SFA, 2008). Pine moth (*Dendrolimus* spp.) infest up to 3.30 million ha across China (Luo, 2002). Furthermore, the damaged area of poplar (*Populus* spp.) pests reached up to 2.67 million (Luo, 2002). In 2010, 11.64 million ha of forests were attacked by pests and diseases, and 7,723 forest fire events took place in planted forests (SFA 2011).

## 5.3 Dealing with limitations of land availability for new planted forests

Further expansion of planted forests in China is increasingly constrained by land availability due to the limited land resources and the continuous competition with other land uses, such as agriculture, settlement and infrastructure development. At present, about 30 million ha of land area potentially available for planted forests suffers from poor climate and soil conditions and it would be a technologically and economically challenging task to bring them under tree-growing regimes (SFA, 2014). In particular, there is less land available for planted forests in eastern regions compared with western areas. Although land is available especially in north-western areas, water resources are very scarce there. Developing planted forests in water-limited areas where the annual precipitation is below 400 mm would potentially exacerbate water shortages due to the increased evapotranspiration, especially under climate-change scenarios. Some studies have shown that the vegetation composition in terms of forest, shrub and alpine meadows could affect the amount of water yield in a watershed (Jiang et al., 2004; Y. Liu et al., 2006) and that the annual mean runoff coefficient and evapotranspiration (ET) may be closely related to landscape structure of watersheds (Li et al., 2006; Jiang et al.,

2004; Y. Liu et al., 2006, 2008). At a macro-scale, it is important to consider the natural climatic constraints of water availability as well as the appropriate choice of species/varieties and associated forest management options that are suitable for water conservation. At the meso-scale, the proportion of native forest plays an important role in the reduction and regulation of water use, and therefore a system of mosaic landscape management may be able to stabilise water flow across plantation landscapes (Ferraz et al., 2012).

# 6. FUTURE PERSPECTIVES

China's National Forestry Strategy sets a goal of further expanding 40 million ha of planted forests in 2020 against the baseline of 2005 (SFA, 2009a). However, there are a number of important challenges to overcome, including: the largely monoculture nature of the plantations with only a few tree species dominating the planted landscape; uneven spatial distribution; undesirable age-class distribution; relatively low volume in growing stock; coupled with the increasing complexity of managing planted forests for multiple uses under a changing environment. Changing the management strategy of planted forests from one of area expansion to one of productivity and quality enhancement cannot come too soon. This change would occur by transforming forest management regimes from those relying solely on massive artificial regeneration and afforestation to ones focusing more on tending and natural regeneration. Such a change would enable the forest managers to contribute to climate mitigation and adaptation, water and biodiversity conservation by implementing management plans that comprehensively integrate economic (productivity and growth), social (equity of access to water and land-use conflicts) and environmental (climate change and biodiversity impacts) benefits from planted forests.

# 7. CONCLUSIONS

Planted forests are increasingly playing an important role in China for wood and non-wood forest products, carbon sequestration, soil stabilisation, and a range of other ecosystem services. Sustainability of planted forests in terms of desired goods and ecosystem services, and diversified adaptive options for different stakeholders to cope with global change and land use conflicts is a growing concern. Understanding of the short-term and long-term impacts of climate change on trees and forests is of great importance to identify the adaptation potentials, and to find ways of improving vitality and resilience of planted forests. There is a need to undertake a wide range of adaptation actions at different management levels that seek to appropriately modify rotation length, tree species, stand structure and growth and yield models, and at the genetic level that prepare the planted forests to benefit from the pending changes in the climate. Adaptation and mitigation must be considered in parallel and should be integrated together into current management regime of planted forests. Recognition of the interrelationships among land use and balancing of trade-off among multiple ecosystem services should underlie the management of planted forest landscapes. There are considerable uncertainties about potential impacts of climate change on planted forests and this knowledge gap needs to be narrowed down through focused research in a wide range of ecological regions.

# AUTHORS' INFORMATION

Shirong Liu, Professor in forest ecology and vice president of the Chinese Academy of Forestry; Shuirong Wu, Professor in forest resources and environmental economics, and forest policy; and Hui Wang, Associate Professor in forest ecology and soil science.

## ACKNOWLEDGEMENTS

This study was funded by the Ministry of Science and Technology (2012BAD22B01), the Ministry of Finance (201404201) and the National Natural Science Fund Committee (31270681). We gratefully acknowledged the cordial invitation and the support from the organising committee of the Third International Congress on Planted Forests (ICPF) for the authors to attend and make a keynote presentation for the Asian perspective in Session 1 "Global Perspective on Planted Forest Development" in the plenary meeting of ICPF, held in Estoril, Portugal, on May 20-21, 2013. We thank Dr Promode Kant for his valuable comments and help with the English.

## COMPETING INTERESTS

The authors declare that they have no competing interests.

## AUTHORS' CONTRIBUTIONS

Shirong Liu designed the study, participated in drafting the manuscript and significantly improved the writing of the whole manuscript; Shuirong Wu participated in the study design, performed the data collection and analysis, and drafted manuscript on the social and economic aspects; Hui Wang participated in the study design and drafted the manuscript on the ecological aspects. All authors read and approved the final manuscript.

# REFERENCES

1. Booth TH: **Eucalypt plantations and climate change.** *Forest Ecology and Management* 2013, **301:** 28–34.

2. Carle J, Holmgren P: **Wood from planted forests: a global outlook 2005–2030.** *Forest Products Journal* 2008,**58**(12):6–18.

3. Cossalter C, Barr C: **Fast-growing plantation development and industrial wood demand in China's Guangxi Zhuang Autonomous Region.** *(A Report Prepared for the Guangxi Forestry Bureau and the World Bank) Bogor, Indonesia: Center for International Forestry Research* 2005.

4. Fang JY, Chen AP, Peng CH, Zhao SQ, Ci L: **Changes in forest biomass carbon storage in China between 1949 and 1998.***Science* 2001, **292:** 2320–2322.

5. Fang JY, Guo ZD, Piao SL, Chen AP: **Terrestrial vegetation carbon sinks in China, 1981–2000.** *Science in China(D-Earth Science)* 2007, **50:** 1341–1350.

6. FAO: **Global planted forest thematic study: results and analysis.** 2006a.

7. FAO: **Responsible management of planted forests: voluntary guidelines.** 2006b.

8. FAO: **Global Forest Resources Assessment 2010.** 2010.

9. Ferraz SFB, Lima WP, Rodrigues CB: **Managing forest plantation landscapes for water conservation.** *Forest Ecology and Management* 2012, **301:** 58–66.

10. General Bureau of State Forestry: *Statistics on the national forest resources (1973–1976).* (In Chinese). Beijing, P. R. China: General Bureau of State Forestry; 1977.

11. Jiang H, Liu SR, Sun PS, An SQ, Zhou GY, Li CY, Wang JX, Yu H, Tian XJ: **The influence of vegetation type on the hydrological process**

**at the landscape scale.** *Canadian Journal of Remote Sensing* 2004,**30**(5):743–763.

12. Kröger M: **Global tree plantation expansion: a review.** *(ICAS Review Paper Series No 3) The Netherlands: Initiatives in Critical Agrarian Studies, Land Deal Politics Initiative and Transnational Institute* 2012. [https://helda.helsinki.fi/bitstream/handle/10138/39217/Markus_Kroeger_ICAS_WP_3_EN.pdf?sequence=2]

13. Lan Y, Gu W: **Advances in trees asexual reproduction.** *World Forestry Research* 2002, **6:** 25–30. (in Chinese)

14. Lane M, McDonald G: **Towards a general model of forest management through time: evidence from Australia, USA, and Canada.** *Land Use Policy* 2002, **19:** 193–206.

15. Li CW, Ge JP, Liu SR, Sun PS: **Landscape pattern and eco-hydrological characteristics at the upstream of Minjiang River, China.** *Frontiers of Biology in China* 2006,**1**(4):455–462.

16. Liu S, Yong L, Zhang Y, Guo Z, Zhang L, Li C, Wang J: **Landscape ecology contributions to forestry and forest management in China: progresses and research needs.** In *Landscape ecology in forest management and conservation: challenges and solutions for global change*. Edited by: C Li, R Lafortezza, & J Chen,. Beijing, P. R. China: Springer, Higher Education Press; 2011:22–45.

17. Liu YH, An SQ, Deng Z, Fan N, Wang Z, Zhou C, Liu S: **Effects of vegetation patterns on yields of the surface and subsurface waters in the Heishui Alpine Valley in west China.** *Hydrology and Earth System Sciences Discussions* 2006, **3:**1021–1043.

18. Liu YH, Fan NJ, An SQ, Bai XH, Liu FD, Xu Z, Wang ZS, Liu SR: **Characteristics of water isotopes and hydrograph separation during the wet season in the Heishui River, China.** *Journal of Hydrology* 2008, **353:** 314–321.

19. Luo XH: **Plantations and the sustainable development of Chinese forestry.** *Journal of Fujian Forestry Science and Technology* 2002,**29**(2):69–71.

20. Minshal GW: **Stream ecosystem theory: A global perspective.** *Journal of the North American Benthological Society*1988,**7**(4):263–288.

21. Ministry of Forestry: *Statistics on the national forest resources (1977–1981).* (In Chinese). Beijing, P. R. China: Ministry of Forestry; 1983.

22. Ministry of Forestry: *Statistics on the national forest resources (1984–1988).* (In Chinese). Beijing, P. R. China: Ministry of Forestry; 1989.

23. Ministry of Forestry: *Statistics on the national forest resources (1989–1993).* (In Chinese). Beijing, P. R. China: Ministry of Forestry; 1994.

24. Niu X, Wang B, Liu S, Liu C, Wei W, Kauppi PE: **Economical assessment of forest ecosystem services in China: Characteristics and implications.** *Ecological Complexity* 2012, **11:** 1–11.

25. Pan YD, Birdsey RA, Fang JY, Houghton RA, Kauppi PE, Kurz WA, Phillips OL, Shvidenko A, Lewis SL, Canadell JG, Ciais P, Jackson RB, Pacala SW, McGuire AD, Piao SL, Rautiainen A, Sitch S, Hayes D: **A large and persistent carbon sink in the world's forests.** *Science* 2011, **333:** 988–993.

26. Peng SL, Wang DX, Zhao H, Yang T: **Discussion the status quality of plantation and near nature forestry management in China.** *Journal of Northwest Forestry University* 2008, **23:** 184–188.

27. Piao S, Fang J, Ciais P, Peylin P, Huang Y, Sitch S, Wang T: **The carbon balance of terrestrial ecosystems in China.** *Nature*2009, **458:** 1009–1013. 23 April

28. Poff LN: **A hydrogeography of unregulated streams in the United States and an examination of scale-dependence in some hydrological descriptors.** *Freshwater Biology* 1996, **36:** 71–91.

29. Poff LN, Ward JV: **Implications of streamflow variability and predictability for lotic community structure: A regional analysis of streamflow patterns.** *Canadian Journal of Fisheries and Aquatic Sciences* 1989, **46:** 1805–1817.

30. Rudel K: **Tree farms: Driving forces and regional patterns in the global expansion of forest plantations.** *Land Use Policy*2009, **26:** 545–550.

31. Sedjo R: *Global supply and demand for forest products: some important trends.* Washington, DC: Resources for the Future; 2004.

32. Shao Q, Huang L, Liu J, Kuang W, Li J: **Analysis of forest damage caused by the snow and ice chaos along a transect across southern China in Spring 2008.** *Journal of Geography Science* 2011,**21**(2):219–234.

33. Sohngen B, Sedjo R: *The potential role of plantations in future timber supply.* Washington, DC: Resources for the Future; 1999.

34. State Forestry Administration (SFA): *Statistics on the national forest resources (1994–1998).* (In Chinese). Beijing, P. R. China: SFA; 2000.

35. State Forestry Administration (SFA): *Bulletin of Chinese forest resources in 2005.* (In Chinese). Beijing, P. R. China: China Forestry Publishing House; 2005.

36. State Forestry Administration (SFA): *Overview on Forest Pests and Disease in China.* (In Chinese). Beijing, P. R. China: China Forestry Publishing House; 2008.

37. State Forestry Administration (SFA): *The Forestry Action Plan to Address Climate Change.* (In Chinese). Beijing, P. R. China: China Forestry Publishing House; 2009a.

38. State Forestry Administration (SFA): *Report on Forest Resources of China - based on the Seventh National Forest Inventory.* (In Chinese). Beijing, P. R. China: China Forestry Publishing House; 2009b.

39. State Forestry Administration (SFA): *Technical regulations for inventory for forest management planning and design (GB/T 26424–2010).* (In Chinese). Beijing, P. R. China: China Standards Press; 2011a.

40. State Forestry Administration (SFA): *China Forestry Statistical Yearbook 2011.* (In Chinese).Beijing, P. R. China: China Forestry Publishing House; 2011b.

41. State Forestry Administration (SFA): *China Forestry Statistical Yearbook 2013.* In Chinese. Beijing, P. R. China: China Forestry Publishing House; 2013a.

42. State Forestry Administration (SFA): *Report on Chinese Forestry Development 2013.* (In Chinese). Beijing, P. R. China: China Forestry Publishing House; 2013b.

43. State Forestry Administration (SFA): **General Situation of Forest Resources in China- based on the Eighth National Forest Inventory.** 2014.

44. Sun PS, Liu SR, Jiang H, Lu YL, Liu JT, Lin Y, Liu XL: **Hydrologic effects of NDVI time series in a context of climatic variability.** *Journal of the American Water Resources Association* 2008,**44**(5):1132–1143.

45. Wang W, Wang WJ, Li JS, Wu H, Xu C, Liu T: **The impact of sustained drought on vegetation ecosystem in Southwest China based on remote sensing.** *Procedia Environmental Sciences* 2010, **2:** 1679–1691.

46. Wijewardana D: **Markets perspectives for timber from planted forests.** *New Zealnd Journal of Forestry* 2005, 34–41. May

47. Winjum JK, Schroeder PE: **Forest plantations of the world: their extent, ecological attributes, and carbon storage.***Agricultural and Forest Meteorology* 1997, **84:** 153–167.

48. Xiao X: *National forest inventory of China.* (In Chinese). Beijing, P. R. China: China Forestry Publishing House; 2005.

49. Zhang Y, Tachibana S, Nagata S: **Impact of socio-economic factors on the changes in forest areas in China.** *Forest Policy and Economics* 2006, **9:** 63–76.

9

# EVALUATION OF MAJOR FACTORS INFLUENCING THE GEOCHEMISTRY OF GROUNDWATER USING GRAPHICAL AND MULTIVARIATE STATISTICAL METHODS IN YENAGOA CITY, SOUTHERN NIGERIA

**K. S. Okiongbo[1], R. K. Douglas[2]**

[1] Department of Geology and Physics, Niger Delta University, Wilberforce Island, Bayelsa State, Nigeria

[2] School of Applied Sciences, Cranfield University, Cranfield, Bedforshire MK 43 0AL, UK

## ABSTRACT

To achieve a better understanding of the nature of the factors influencing groundwater composition as well as to specify them quantitatively, conventional graphical and multivariate statistical analysis (principal component analysis) were applied on hydrochemical data consisting of 51 groundwater samples collected from domestic boreholes in Yenagoa city, Bayelsa State, Nigeria. The mode of study includes analysis of major ion contents and other chemical parameters such as pH, total dissolved solids and electrical conductivity of the groundwater samples. The PCA yielded three principal components explaining 78.38 % of the total variance of the 11 parameters. The three components are interpreted as controlled by the natural weathering of existing silicate rocks, reverse ion-exchange processes and oxidation reactions which are further supported by the scatter diagrams, ionic signatures and mechanisms controlling the water chemistry diagrams as the common factors influencing the groundwater hydrogeochemical character. Limited anthropogenic influence on the groundwater composition has also been noticed in the study area. The groundwater poses no threat to human health because the concentrations of physico-chemical parameters that can be used to evaluate drinking water quality are within World Health Organisation standard specification. The groundwater in the area is fresh, high salinity and low sodium in nature.

## KEYWORDS

Multivariate statistical analysis, R-mode factor analysis, Groundwater quality, Hydrogeochemistry, Yenagoa

## 1. INTRODUCTION

Hydrogeochemical investigations to characterise groundwater and assess its chemical quality often involve substantially large amount of base information concerning groundwater chemistry. Groundwater chemistry, in turn, depends on a large number of factors, such as general geology, degree of chemical weathering of the various rock types, quality of recharge water and inputs from sources other than water–rock interaction. Such factors and their interactions result in a complex groundwater quality. Thus, the characterisation and interpretation of various groundwater parameters is often a complex problem. Several conventional methods of data analysis are often used for the interpretation and presentation of results (Hem 1989), e.g. histograms, stiff, trilinear, semilogarithmic diagrams and many others (Lloyd and Heathcote 1985). The use of these graphical methods to interpret aquatic chemistry is limited to two dimensions, besides, deal only with a limited number of variables responsible for the groundwater chemistry and can produce biased results (Guler et al. 2002).

To overcome the limitations of these conventional methods, multivariate statistical analysis such as principal component analysis had been widely used for characterizing and interpreting geochemical data. Such analyses are especially useful because in addition to revealing the interaction among the variables, it also attempts to simplify the complex and diverse relationships that exist among a set of observed variables (i.e. the concentrations of physico-chemical constituents) by revealing common and unobserved factors (i.e. the underlying physico-chemical constituents and/or biological processes in the groundwater system) that link together with the seemingly unrelated variables (Lee et al. 2001). Reeder et al. (1972) identified the likely weathering processes controlling the chemical composition of surface waters of Mackenzie River drainage basin Canada. Ashley and Lloyd (1978) used factor analysis to evaluate the hydrogeochemical process in the Santiago basin of Chile and the Derbyshire Dome of England. Lawrence and Upchurch (1982) also used the technique to delineate the zones of natural recharge to groundwater in the Florida aquifer. Ruiz et al. (1990) suggested that the basic purpose of such an analysis was to study the hydrogeochemistry of aquifers by simplifying the numerous and complex groundwater data into a set of factors, few in number which can explain a large amount of the variance of the analytical data and also indicate the source of origin of various ions present in water. Earlier studies carried out in the study area are on the geochemistry of groundwater (Amadi et al. 1987), prospecting of groundwater resources (Okiongbo and Ogobiri 2011), and groundwater quality with respect to drinking and agricultural purposes (Okiongbo and Douglas 2013). These studies mainly focus on chemical composition of groundwater using conventional methods only.

In this study, we carry out a hydrochemical evaluation of an alluvial aquifer system with an objective of delineating the various factors controlling the water chemistry and general suitability of the groundwater for domestic and drinking purposes using conventional graphical and multivariate statistical methods.

# 2. DESCRIPTION OF THE STUDY AREA

## 2.1 Location and climate

The study area entails Yenagoa city, part of Southern Ijaw and Ogbia Local Government Areas in Bayelsa State and lies between longitudes 006°05′ and 00 6°040′ East of the prime meridian and latitudes 04°23.3′ and 04°38.2′ North of the equator within the coastal area of the recent Niger Delta (Fig. 1). Bayelsa State is located in the southern end of Nigeria and is within the Niger Delta Basin. It is bounded in the north by Rivers State, in the west by Delta State and in the south by the Bight of Bonny. The study area has an areal extent of about 150 km². The study area has a tropical rain forest climate characterised by two seasons, namely the wet or rainy season and the dry season. The rainy season lasts for about 7 months between April and October with an intervening dry period in August. The dry season lasts for about 4 months, between November and March. The temperature varies between 25 and 32 °C. The mean annual rainfall is about 4,500 mm (Akpokodje 1986); about 85 % of the mean annual rain falls in the wet season.

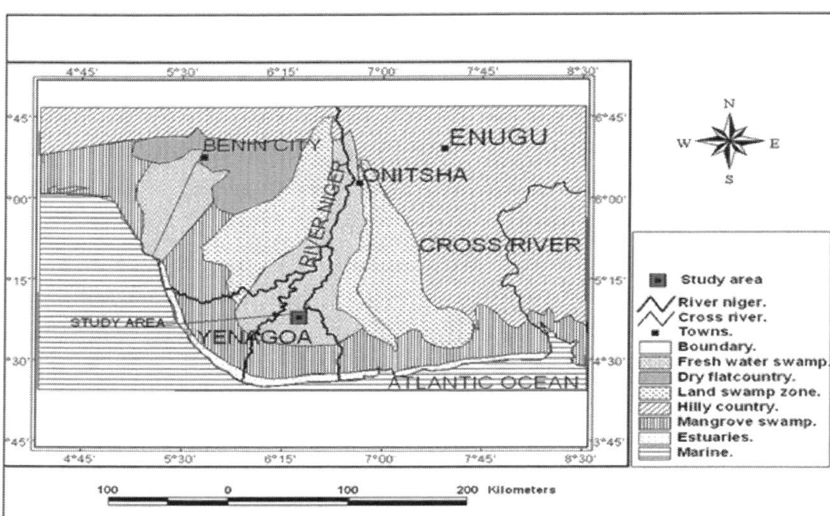


**Figure 1.**The major geomorphological units of the Niger Delta (After Akpokodje 1986)

## 2.2 Physiography and drainage

The study area consists of alluvial deposits and an extensive, low-lying, typical deltaic plain with essentially flat topography which, in conjunction with the high annual rainfall, is responsible for the extremely poor drainage conditions and the widespread development of marshes and backswamps. This area is usually submerged during the wet season where flood waters range from 0.5 to 4 m deep (Akpokodje 1986). There are a number of perennial streams, oxbow lakes and rivers in the area e.g. Kolo Creek, Epie Creek, Yenagoa and Nun river, etc. They all form a network which empties to the Atlantic Ocean through Nun River Estuary. These rivers are mostly turbid during the wet season possibly due to discharge of clay and silt (Amadi et al.1987). The natural vegetation of the study area is that of the rain forest but this has been destroyed by the activities of man such as bush burning, farming, construction and illegal crude oil refining activities. The vegetation consists of various kinds of evergreen trees, including palms trees and a variety of shrubs. More than 70 % of the inhabitants of the study area are engaged in subsistent farming and fishing.

## 2.3 Geology

The study area (Fig. 1) lies within the fresh water forests, swamps and backswamp geomorphic unit of the Niger Delta. The Niger Delta is basically an alluvial plain formed by the deposition and built-up of fine-grained sediments eroded and transported to the area by the River Niger and its tributaries (Fig. 1). The regional geology of the Niger Delta is relatively simple consisting of three lithostratigraphic units, Akata, Agbada and Benin Formations, overlain by various types of quaternary deposits (Short and Stauble 1967; Wright et al. 1985; Kogbe 1989).

The quaternary deposits are considered universally to be the recent expressions of and continuation of the Benin Formation and consist of medium to coarse-grained sands, sandy clays, silts and subordinate, lensoid clay bands thought to have been deposited during quaternary interglacial marine transgressions (Durotoye1989). Amajor (1991) has shown that they are an admixture of fluvial/tidal channel, tidal flats and mangrove swamp deposits. The sands are micaceous and feldspathic, subrounded to angular in texture and constitute good aquifers. However, depth to occurrence and thickness is irregular and may not be predicted with accuracy within the study area due to rapid horizontal and vertical facie changes. Groundwater occurs generally under water table conditions where the lensoid clays attain adequate thickness.

The Benin Formation is essentially fluvial in origin comprising unconsolidated, massive and porous fresh water bearing sands with localized clay/shale beds. The sand-clay intercalations are indicative of a multi-aquifer system. The thickness of the Benin Formation is about 2,100 m on the average. The age of this Formation ranges from Miocene to Recent. Onyeagocha (1980) reported that the rocks are made up of 95 % quartz grains, Na + mica 1–2.5 %, feldspar 0–1.0 % and dark coloured minerals 2.3 %. Water table in large sections of the Niger Delta is close to the surface but subject to spatial and seasonal variations. In the study area, the water table is about 3–4 m during the dry season. During the wet season, the water table rises considerably, in some cases, to the ground surface. The Agbada Formation underlies the Benin Formation. Its thickness varies from 300 to 4,500 m (Short and Stauble 1967). This Formation consists predominantly of unconsolidated pebbles, and very coarse to fine-grained sand units with subordinate shale beds. Underneath the Agbada Formation is the Akata Formation of Eocene to recent age (Amadi et al. 1987), which is made up of shales with local inter-bedding of sands and siltstones. It was deposited in a typically marine environment with maximum thickness of about 6,100 m (Ofoegbu 1989).

## 2.4 Hydrogeology

The Coastal Plain Sands of the Benin Formation are the main regional and most important aquifer in the study area. Groundwater in the Coastal Plain Sands occurs mainly under phreatic (unconfined) conditions. The lithology of this formation is dominated by loose sands (fine-medium-coarse), while gravel and pebbles are minor components. Thin clay horizons and lenses create discontinuities in the vertical and lateral continuity of the porous sands and gravel. This condition results in the presence of local perched aquifers. Rainfall is the direct recharge source of the groundwater. The transmissivity values for the aquifer range from $1.05 \times 10^{-2}$ to $11.3 \times 10^{-2} m^2/s$, while the coefficient of storage varies between $1.07 \times 10^{-4}$ and $3.53 \times 10^{-4}$and specific capacity values lie between 19.01 and 139.8 $m^3/h/m$ drawdown (Amajor 1991). These values suggest that the aquifers have very good capacity to transmit groundwater. The groundwater quality is very good and compares favourably with WHO standards for drinking water. However, relatively high iron and chloride values though localized are observed.

## 3. METHOD OF STUDY

Groundwater samples were collected from 51 domestic boreholes spread over the study area (Fig. 2) in 3 weeks in the month of February, 2013, and were analysed within 24 h. A global positioning system (GPS) was used for location and elevation readings. The samples were collected in polyethylene bottles for the analysis of total hardness, total dissolved solids (TDS) and major ions such as Na, K, Ca, Mg, $HCO_3$, $CO_3$, Cl and $SO_4$ in the laboratory. All samples were analysed using standard procedures (APHA 1995). Electrical conductivity (EC) and pH of groundwater samples were measured in the field immediately after sample collection using portable field kits (Hach's CO 150 Conductivity/TDS metre and pH metre, respectively). Major ions like sulphate ($SO_4^{2-}$) were determined by spectrophotometric turbidimetry. The method is based on the principle of formation of barium sulphate in the presence of (acidified HCl) barium chloride. The process is enhanced in the presence of glycerol or other organic compound. The absorbance of the colloidal solution was measured against a standard on UV-visible Spectrophotometer. Calcium ($Ca^{2+}$) and magnesium ($Mg^{2+}$) were determined titrimetrically using EDTA; chloride ($Cl^-$) was determined by standard $AgNO_3$ titration; bicarbonate ($HCO_3^-$) was determined by titration with HCL; sodium ($Na^+$) and potassium ($K^+$) were measured by flame photometry; nitrate ($NO_3^-$) was determined by colourimetry with a UV-visible spectrophotometer (brucine method) while iron was determined by colourimetry with a UV-visible spectrophotometer at 520 nm. Iron results represent ferrous iron. In general, ion balance error of the samples was observed to be within the acceptable range of ±10 %.

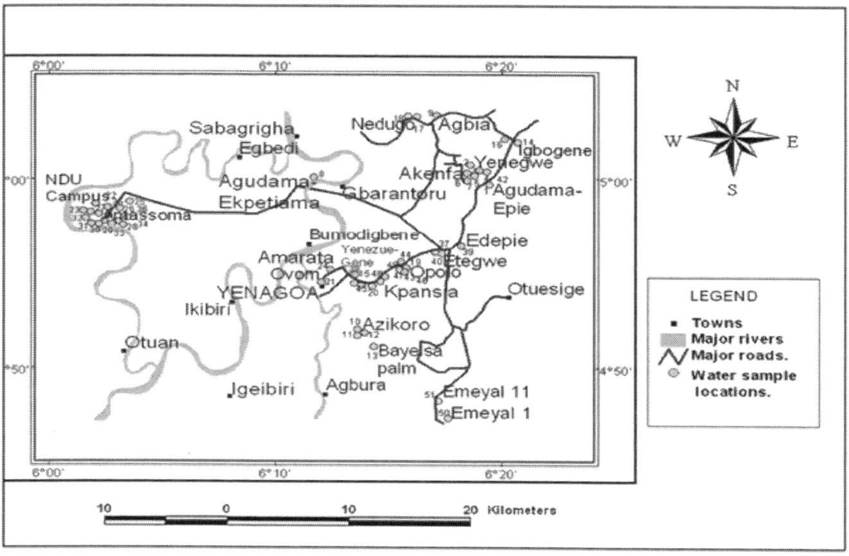


**Figure 2.** Map of study area showing borehole locations

The data was analysed using multivariate statistical technique (principal component analysis) to identify the hydrogeochemical processes which result in the variations in the chemical composition of groundwater. Saturation indices of groundwater with respect to the mineral phases and the chemical facies were determined using the computer programmes PHREEQC and AQUACHEM, respectively.

The obtained matrix of hydrogeochemical data was subjected to multivariate analytical technique. Principal component analysis (PCA) was used here as a numerical method of discovering variables that are more important than others for representing parameter variation and identifying hydrogeochemical processes. The entire dataset was first standardized and arranged in correlation coefficient matrix with normal distribution in all variables. The principal component approach started by extracting eigenvalues and eigenvectors of the correlation matrix and then less important of these was discarded (Davis 1986). The eigenvalues calculated quantify the contribution of a factor to the total variance. The percentages were also computed. The contribution of a factor is significant when the eigenvalue is greater than unity. The factors that are significant (with eigenvalues more than one and explaining for the higher percentage of the variability of the data) were noted. The factor loadings were calculated by a varimax rotation technique in such a way that they are closer to +1, 0, −1, representing positive contribution, no contribution and negative contribution. They suggest the dominance of the corresponding variables which, when projected onto the data, give factor scores.

# 4. RESULTS AND DISCUSSION

## 4.1 Groundwater chemistry

The physico-chemical compositions of the groundwater samples are given in Table 1, and Table 2 gives the descriptive statistical measures such as minimum, maximum, mean and standard deviation values of various parameters investigated along with the World Health Organisation standards (WHO 2004). In the groundwater samples, pH value ranges from 5.55 to 6.93 with an average of 6.93. The decrease in pH is primarily due to oxidation of organic matter and also due to anthropogenic activities (Krishnaraj et al. 2011). Concentrations of TDS, a measure of groundwater quality ranged from 51 to 1,410 mg/L with an average of 388.3 mg/L. According to TDS classification, 90 % of the samples belong to fresh water type (TDS <1,000 mg/L) and the remaining 10 % represent brackish water (TDS > 1,000 mg/L). The electrical conductivity values range from 102 to 2,820 μS/cm, with an average of 765.9 μS/cm.

**Table 1.** Hydrogeochemical analysis of groundwater samples

| S/N | Code | pH | EC | TDS | TA | TH | $Na^+$ | $K^+$ | $Ca^{2+}$ | $Mg^{2+}$ | Fe | $Cl^-$ | $SO_4^{2-}$ | $NO_3^-$ | $HCO_3$ | CAI-1 | CA1-2 | SO4/Cl | SAR |
|---|---|---|---|---|---|---|---|---|---|---|---|---|---|---|---|---|---|---|---|
| 1 | BH1 | 6.83 | 2,120.0 | 1,060.0 | 55.0 | 143.0 | 52.2 | 30.4 | 40.5 | 10.7 | 0.50 | 31.8 | 7.2 | 0.13 | 300.8 | −2.4 | −0.4 | 0.23 | 2.7 |
| 2 | BH2 | 6.16 | 338.0 | 169.0 | 15.0 | 48.0 | 14.2 | 6.8 | 2.4 | 0.5 | 0.02 | 2.9 | 0.5 | 0.11 | 52.5 | −8.9 | −0.8 | 0.18 | 3.1 |
| 3 | BH3 | 5.78 | 615.0 | 308.0 | 26.0 | 25.0 | 28.4 | 16.5 | 4.0 | 0.8 | 0.01 | 4.8 | 4.2 | 0.08 | 103.7 | −11.2 | −0.9 | 0.87 | 4.8 |
| 4 | BH4 | 5.87 | 472.0 | 236.0 | 16.0 | 16.0 | 14.5 | 12.0 | 6.1 | 3.5 | 0.18 | 4.6 | 1.2 | 0.08 | 84.2 | −6.2 | −0.6 | 0.26 | 1.6 |
| 5 | BH5 | 6.17 | 400.0 | 200.0 | 20.0 | 6.0 | 18.2 | 9.9 | 5.4 | 3.1 | 0.15 | 12.5 | 1.8 | 0.04 | 72.0 | −2.0 | −0.6 | 0.14 | 2.2 |
| 6 | BH6 | 5.82 | 768.0 | 384.0 | 39.0 | 50.0 | 16.5 | 6.8 | 5.5 | 2.5 | 0.24 | 5.5 | 4.6 | 0.27 | 68.3 | −4.7 | −0.6 | 0.83 | 2.1 |
| 7 | BH7 | 6.03 | 515.0 | 256.0 | 21.0 | 44.0 | 18.4 | 7.5 | 3.5 | 1.2 | 0.18 | 4.8 | 1.7 | 0.16 | 67.1 | −6.3 | −0.8 | 0.35 | 3.1 |
| 8 | BH8 | 6.38 | 708.0 | 354.0 | 30.0 | 48.0 | 28.4 | 9.5 | 6.8 | 3.4 | 0.12 | 17.3 | 2.8 | 0.03 | 94.6 | −2.0 | −0.6 | 0.16 | 3.1 |
| 9 | BH9 | 6.20 | 2,620.0 | 1,310.0 | 48.0 | 254.0 | 60.4 | 28.5 | 36.4 | 8.5 | 0.02 | 82.0 | 5.7 | 0.13 | 209.8 | −0.5 | −0.3 | 0.07 | 3.3 |
| 10 | BH10 | 6.24 | 800.0 | 400.0 | 28.0 | 40.0 | 14.5 | 8.7 | 5.6 | 3.5 | 0.25 | 11.3 | 0.9 | 0.16 | 66.5 | −1.7 | −0.5 | 0.08 | 1.7 |
| 11 | BH11 | 6.15 | 1,036.0 | 518.0 | 24.0 | 40.0 | 29.5 | 18.4 | 6.3 | 2.7 | 0.05 | 14.4 | 3.2 | 0.14 | 111.0 | −3.3 | −0.7 | 0.22 | 3.5 |
| 12 | BH12 | 5.92 | 907.0 | 454.0 | 42.0 | 57.0 | 25.5 | 15.6 | 8.5 | 2.2 | 0.29 | 14.8 | 4.2 | 0.05 | 98.2 | −2.6 | −0.6 | 0.28 | 2.9 |
| 13 | BH13 | 6.16 | 696.0 | 348.0 | 42.0 | 52.0 | 12.5 | 8.2 | 4.5 | 5.6 | 0.32 | 14.5 | 2.5 | 0.01 | 60.4 | −0.8 | −0.3 | 0.17 | 1.3 |
| 14 | BH14 | 5.80 | 1,091.0 | 546.0 | 33.0 | 120.0 | 42.2 | 17.6 | 5.5 | 2.3 | 0.18 | 21.2 | 4.0 | 0.09 | 126.9 | −2.8 | −0.8 | 0.19 | 5.4 |
| 15 | BH15 | 6.31 | 418.0 | 209.0 | 30.0 | 15.0 | 12.2 | 10.5 | 6.74 | 3.46 | 0.25 | 16.0 | 0.54 | 0.09 | 59.2 | −0.8 | −0.4 | 0.03 | 1.3 |
| 16 | BH16 | 5.87 | 1,324.0 | 662.0 | 60.0 | 54.0 | 41.7 | 18.4 | 12.8 | 8.8 | 0.21 | 20.4 | 3.5 | 0.14 | 183.0 | −3.0 | −0.6 | 0.17 | 3.1 |
| 17 | BH17 | 5.87 | 1,098.0 | 549.0 | 40.0 | 48.0 | 15.4 | 9.8 | 10.7 | 6.2 | 0.31 | 21.6 | 10.6 | 0.00 | 70.2 | −0.5 | −0.2 | 0.49 | 1.3 |
| 18 | BH18 | 6.00 | 440.0 | 220.0 | 39.0 | 23.5 | 15.2 | 7.4 | 4.6 | 2.5 | 0.38 | 11.7 | 5.5 | 0.28 | 51.9 | −1.6 | −0.5 | 0.47 | 2.0 |
| 19 | BH19 | 6.04 | 207.0 | 104.0 | 13.0 | 14.2 | 3.8 | 2.5 | 2.9 | 1.6 | 0.32 | 10.3 | 2.1 | 0.20 | 11.0 | 0.2 | 0.3 | 0.20 | 0.6 |
| 20 | BH20 | 6.34 | 2,820.0 | 1,410.0 | 65.0 | 265.0 | 48.7 | 15.2 | 20.9 | 6.9 | 0.18 | 39.1 | 6.5 | 0.42 | 175.7 | −1.3 | −0.5 | 0.17 | 3.3 |
| 21 | BH21 | 6.88 | 792.0 | 396.0 | 30.0 | 68.0 | 18.8 | 8.2 | 8.0 | 1.8 | 0.18 | 8.4 | 5.4 | 0.15 | 70.0 | −3.4 | −0.6 | 0.65 | 2.2 |
| 22 | BH22 | 6.18 | 495.0 | 248.0 | 18.0 | 42.0 | 13.8 | 6.5 | 3.5 | 1.8 | 0.86 | 14.2 | 4.7 | 1.60 | 36.0 | −0.9 | −0.5 | 0.33 | 2.1 |
| 23 | BH23 | 6.19 | 409.0 | 205.0 | 32.0 | 23.0 | 17.2 | 8.3 | 5.0 | 4.5 | 0.28 | 6.9 | 2.6 | 0.05 | 81.7 | −3.9 | −0.5 | 0.38 | 1.9 |
| 24 | BH24 | 6.70 | 591.0 | 296.0 | 26.0 | 24.0 | 7.8 | 12.0 | 5.7 | 4.2 | 0.20 | 5.3 | 3.3 | 0.28 | 64.7 | −3.3 | −0.4 | 0.62 | 0.9 |
| 25 | BH25 | 6.06 | 910.00 | 455.00 | 23.00 | 60.00 | 12.40 | 7.60 | 13.50 | 3.80 | 10.10 | 12.50 | 2.30 | 6.74 | 8.40 | −1.08 | −1.15 | 0.18 | 1.1 |

| | | | | | | | | | | | | | | | | | | | |
|---|---|---|---|---|---|---|---|---|---|---|---|---|---|---|---|---|---|---|---|
| 26 | BH26 | 5.80 | 890.00 | 450.00 | 17.00 | 85.00 | 10.60 | 5.80 | 14.80 | 12.60 | 2.08 | 28.40 | 1.20 | 1.68 | 6.20 | 0.24 | 1.17 | 0.04 | 0.7 |
| 27 | BH27 | 6.20 | 1,060.00 | 530.00 | 22.00 | 57.00 | 12.50 | 10.50 | 28.50 | 21.80 | 11.94 | 49.70 | 3.00 | 5.85 | 10.80 | 0.42 | 1.60 | 0.06 | 0.6 |
| 28 | BH28 | 6.22 | 1,200.00 | 600.00 | 26.00 | 75.00 | 9.50 | 12.60 | 18.80 | 6.70 | 12.00 | 17.70 | 5.40 | 8.56 | 18.20 | −0.47 | −0.40 | 0.31 | 0.7 |
| 29 | BH29 | 6.19 | 690.00 | 345.00 | 16.00 | 55.00 | 8.50 | 4.60 | 15.70 | 6.50 | 10.61 | 24.50 | 1.10 | 1.50 | 5.60 | 0.29 | 1.38 | 0.04 | 0.6 |
| 30 | BH30 | 6.44 | 1,163.00 | 582.00 | 26.00 | 65.00 | 12.30 | 6.40 | 16.50 | 10.80 | 6.06 | 21.30 | 6.50 | 2.45 | 14.80 | −0.16 | −0.23 | 0.31 | 0.8 |
| 31 | BH31 | 6.46 | 1,089.00 | 545.00 | 15.00 | 75.00 | 8.80 | 9.20 | 15.20 | 8.00 | 7.38 | 14.20 | 4.80 | 3.40 | 9.40 | −0.54 | −0.66 | 0.34 | 0.6 |
| 32 | BH32 | 6.65 | 1,138.00 | 569.00 | 18.00 | 166.00 | 10.50 | 9.60 | 8.80 | 6.50 | 6.24 | 12.40 | 3.50 | 1.30 | 8.80 | −1.01 | −1.44 | 0.28 | 0.9 |
| 33 | BH33 | 6.04 | 526.00 | 520.00 | 25.00 | 4.00 | 8.40 | 9.60 | 6.80 | 7.20 | 12.00 | 10.50 | 0.50 | 2.70 | 12.70 | −1.06 | −1.14 | 0.05 | 0.8 |
| 34 | BH34 | 6.38 | 1,068.00 | 534.00 | 15.00 | 73.00 | 7.40 | 7.50 | 10.20 | 5.80 | 12.08 | 17.00 | 0.84 | 4.85 | 10.50 | −0.07 | −0.12 | 0.05 | 0.6 |
| 35 | BH35 | 6.84 | 591.00 | 296.00 | 38.00 | 37.00 | 6.50 | 7.80 | 10.80 | 8.20 | 9.08 | 30.40 | 0.38 | 2.20 | 4.00 | 0.44 | 3.08 | 0.01 | 0.5 |
| 36 | BH36 | 6.93 | 593.00 | 297.00 | 10.00 | 15.00 | 6.40 | 7.40 | 2.80 | 6.90 | 12.10 | 15.00 | 0.54 | 3.70 | 3.80 | −0.11 | −0.29 | 0.04 | 0.7 |
| 37 | BH37 | 6.31 | 2,250.00 | 1,125.00 | 42.00 | 260.00 | 38.90 | 14.00 | 15.00 | 5.70 | 12.26 | 18.50 | 12.40 | 4.60 | 40.00 | −2.93 | −1.51 | 0.67 | 3.1 |
| 38 | BH38 | 6.02 | 277.00 | 139.00 | 12.40 | 13.50 | 15.60 | 10.50 | 10.20 | 6.40 | 1.20 | 21.00 | 8.40 | 0.32 | 4.80 | −0.60 | −1.36 | 0.40 | 1.3 |
| 39 | BH39 | 5.72 | 191.00 | 96.00 | 8.60 | 6.80 | 8.50 | 4.60 | 5.40 | 3.20 | 3.40 | 9.80 | 3.80 | 0.12 | 2.80 | −0.76 | −1.65 | 0.39 | 1.0 |
| 40 | BH40 | 5.77 | 508.00 | 254.00 | 14.50 | 15.00 | 17.70 | 14.20 | 12.60 | 6.80 | 0.15 | 19.90 | 15.40 | 0.35 | 8.60 | −1.02 | −1.22 | 0.77 | 1.4 |
| 41 | BH41 | 6.00 | 331.00 | 165.00 | 15.00 | 19.00 | 8.50 | 6.70 | 11.40 | 7.60 | 0.22 | 22.80 | 14.60 | 0.16 | 6.20 | 0.16 | 0.25 | 0.64 | 0.7 |
| 42 | BH42 | 5.78 | 201.00 | 100.00 | 13.50 | 16.70 | 7.50 | 5.40 | 8.60 | 4.80 | 4.60 | 18.20 | 13.60 | 0.15 | 5.80 | 0.09 | 0.13 | 0.75 | 0.7 |
| 43 | BH43 | 5.70 | 323.00 | 162.00 | 14.20 | 9.80 | 26.50 | 12.70 | 6.25 | 3.24 | 5.20 | 22.20 | 16.50 | 0.13 | 6.90 | −1.36 | −1.85 | 0.74 | 3.0 |
| 44 | BH44 | 5.69 | 348.00 | 174.00 | 15.10 | 14.00 | 18.30 | 10.88 | 8.94 | 6.41 | 2.50 | 20.00 | 15.40 | 0.29 | 7.20 | −0.91 | −1.15 | 0.77 | 1.6 |
| 45 | BH45 | 5.71 | 335.00 | 168.00 | 12.00 | 11.70 | 7.80 | 4.10 | 7.52 | 4.70 | 3.40 | 17.70 | 15.30 | 0.20 | 6.70 | 0.11 | 0.13 | 0.86 | 0.8 |
| 46 | BH46 | 5.64 | 119.00 | 60.00 | 2.50 | 3.50 | 3.80 | 3.50 | 3.40 | 2.85 | 4.80 | 16.50 | 9.54 | 0.12 | 3.80 | 0.45 | 0.80 | 0.58 | 0.5 |
| 47 | BH47 | 5.86 | 161.00 | 82.00 | 3.80 | 4.70 | 1.07 | 4.70 | 4.85 | 3.42 | 2.60 | 18.50 | 9.35 | 0.14 | 2.80 | 0.68 | 1.45 | 0.51 | 0.1 |
| 48 | BH48 | 6.16 | 102.00 | 51.00 | 2.00 | 7.50 | 1.42 | 1.02 | 1.20 | 0.85 | 0.38 | 19.10 | 0.85 | 0.15 | 1.60 | 0.84 | 9.54 | 0.04 | 0.3 |
| 49 | BH49 | 5.55 | 294.00 | 147.00 | 13.00 | 12.60 | 4.28 | 1.85 | 3.50 | 2.80 | 0.25 | 12.30 | 12.60 | 0.28 | 5.50 | 0.33 | 0.32 | 1.02 | 0.8 |
| 50 | BH50 | 6.35 | 112.00 | 60.00 | 2.20 | 3.00 | 2.80 | 2.00 | 2.40 | 1.60 | 0.42 | 36.50 | 16.72 | 0.07 | 2.00 | 0.83 | 2.23 | 0.46 | 0.5 |
| 51 | BH51 | 6.33 | 909.00 | 455.00 | 28.50 | 18.60 | 38.50 | 12.60 | 18.50 | 6.70 | 3.50 | 54.80 | 28.80 | 0.28 | 14.50 | −0.29 | −0.54 | 0.53 | 2.8 |

All parameters have been expressed as mg/L, except pH and EC. The unit of EC is µS/cm while total alkalinity (TA) and total hardness (TH) is mg/L $CaCO_3$. The iron results represent ferrous iron and SAR is expressed in meq/l

**Table 2.** Descriptive statistics of analysed groundwater samples compared with standards

| Parameter | No of Samples | Minimum | Maximum | Mean | Standard Deviation | WHO 2004 maximum permissible limit |
|---|---|---|---|---|---|---|
| pH | 51 | 5.55 | 6.93 | 6.13 | 0.34 | 6.5–8.5 |
| EC (µS/cm) | 51 | 102 | 2,820 | 765.9 | 601.78 | 1,400 |
| TDS (mg/L) | 51 | 51 | 1,410 | 388.3 | 300.87 | 1,000 |
| TA (mg/L) | 51 | 2 | 65 | 24.4 | 14.46 | |
| TH (mg/L) | 51 | 3 | 265 | 53 | 62.60 | 500 |
| $Na^+$ (mg/L) | 51 | 1.1 | 60.4 | 17.4 | 13.65 | 200 |
| $K^+$ (mg/L) | 51 | 1.02 | 30.4 | 9.9 | 5.83 | 55 |
| $Ca^{2+}$ (mg/L) | 51 | 1.2 | 40.5 | 9.9 | 8.05 | 75 |
| $Mg^{2+}$ (mg/L) | 51 | 0.5 | 21.8 | 5.2 | 3.66 | 50 |
| Fe (mg/L) | 51 | 0.01 | 12.3 | 3.18 | 4.27 | 0.3 |
| $Cl^-$ (mg/L) | 51 | 2.9 | 54.8 | 19.2 | 13.80 | 250 |
| $SO_4^{2-}$ (mg/L) | 51 | 0.5 | 28.8 | 6.1 | 5.89 | 400 |
| $NO_3^-$ (mg/L) | 51 | 0.01 | 8.56 | 1.12 | 1.94 | 50 |
| $HCO_3^-$ (mg/L) | 51 | 6.2 | 300.8 | 50 | 62.13 | – |

Tables 1 and 2 show that the major cations $Na^+$, $K^+$, $Ca^{2+}$, $Mg^{2+}$ and Fe are all low except iron with mean concentrations of 17.4, 9.9, 9.9, 5.2 and 3.18 mg/L, respectively. The major anions $HCO_3{}^-$, $Cl^-$, $SO_4{}^{2-}$and $NO_3{}^-$ are also low with mean concentrations of 50 mg/L, 19.2 mg/L, 6.1 mg/L and 1.12 mg/L, respectively. The results obtained show that the concentration of dissolved major cations and anions in the groundwater varies spatially and mass abundance was in the order of: $Na^+ > Ca^{2+} > K^+ > Mg^{2+} >$ Fe for cations and $HCO_3{}^- > Cl^- > SO_4{}^{2-} - > NO_3{}^-$ for anions. The average concentrations of the major ions in the groundwater indicate that the groundwater of the area is suitable for drinking in all samples except iron. The analytical data plotted on the EC versus SAR (sodium adsorption ratio = $Na^+/([Ca^{2+} + Mg^{2+}]/2)^{0.5}$ diagram (Fig. 3) illustrates that most of the groundwater samples fall mainly in the field of C2–S1 and C3–S1 indicating medium to high salinity and low sodium water. Thus, the groundwater in the area can be used for irrigation on almost all types of soil with little danger of exchangeable sodium.

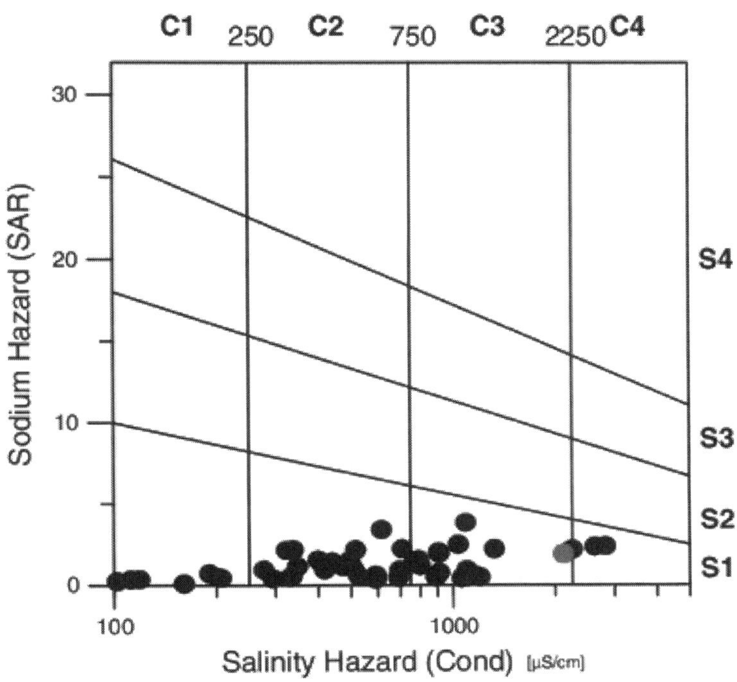


**Figure 3.** Wilcox salinity diagram

A Piper diagram was plotted to examine the hydrochemical water types (Fig. 4). The Piper diagram shows that the chemical composition of the groundwater in the study area is composed of six main types: Ca–Cl, Ca–$HCO_3$–Cl, Ca–Na-Cl, Ca–Na–$HCO_3$–Cl, Ca–Na–$HCO_3$ and Na-$HCO_3$. About 52.2 % groundwater is dominated by the alkalis ($Na^+$ and $K^+$) over the alkaline earths ($Ca^{2+}$ and $Mg^{2+}$) and in 54.2 % of samples, weak acids ($HCO_3{}^-$) exceed strong acids ($SO_4{}^-$ and $Cl^-$).

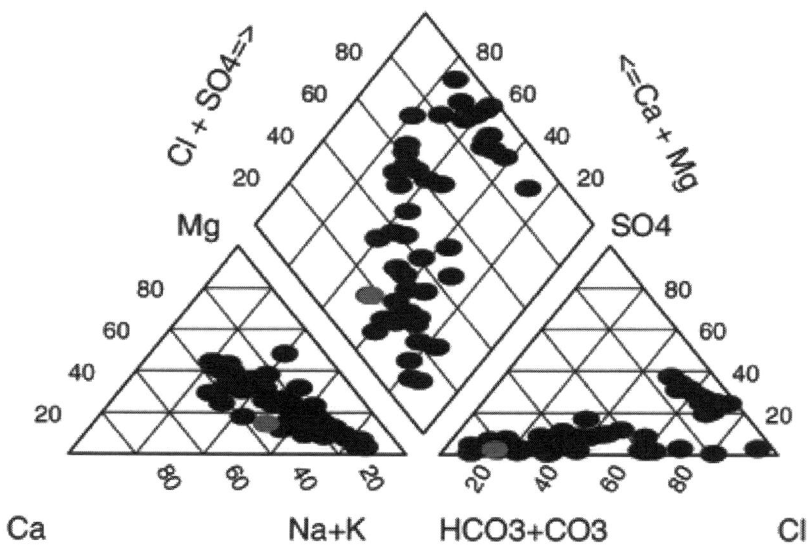


**Figure 4.** Piper trilinear diagram.

## 4.2 Hydrogeochemical evaluation

The geochemical variations in the ionic concentrations in the groundwater can easily be understood when they are plotted along an *X*–*Y* coordinate (Guler et al. 2002). Results from the chemical analyses were used to identify the geochemical processes and mechanisms in the groundwater aquifer system. Gibbs plots of TDS versus the ion ratios Na/(Na + Ca) and Cl/(Cl + $HCO_3$) (Fig. 5) show that all data points fall on the intermediate area between the rock dominance end member and evaporation/precipitation dominance end member indicating that rock weathering was the dominant process controlling the major ion composition of groundwater in the study area (Gibbs 1970).

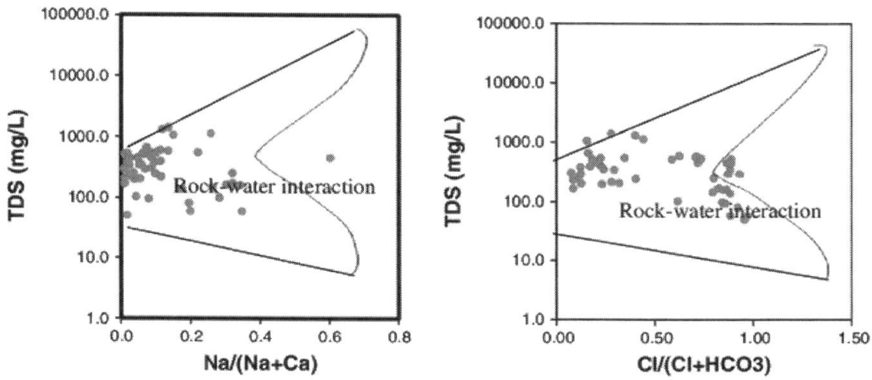


**Figure 5.** Relation Na/(Na + Ca) vs TDS and between Cl/(Cl + $HCO_3$) vs TDS diagrams indicating water–rock interaction

The molar ratio of Na/Cl for groundwater samples of the study area generally ranges from 0.11 to 9.09 (Fig. 6a). Samples having Na/Cl ratio greater than one indicate excess sodium, which might have come from silicate weathering which is dominant in the aquifer materials of the study area. If silicate weathering is a probable source of sodium, the groundwater samples would have $HCO_3^-$ as the most dominant abundant anion (Rogers 1989). This is because the reaction of feldspar minerals with carbonic acid in the presence of water releases $HCO_3^-$ (Elango et al. 2003). Table 1 shows that $HCO_3^-$ is the dominant anion in the groundwater of the study area; hence silicate weathering was the primary process responsible for the release of $Na^+$ into the groundwater. However, samples with a Na/Cl ratio around and less than one indicate the possibility of some other chemical processes, such as ion exchange (Elango et al. 2003).

Cerling et al. (1989) and Fisher and Mulican (1997) observed that the plot of $Ca^{2+} + Mg^{2+}$ vs $SO_4 + HCO_3^-$ is a major indicator to identify ion-exchange process. If ion exchange is the process, the points tend to shift to the right side of the plot due to an excess of $SO_4$ + $HCO_3$ $^-$. If reverse ion exchange is the process, points tend to shift to the left side due to excess of Ca + Mg over $SO_4$ + $HCO_3$ $^-$. The plot of the $Ca^{2+}$ + $Mg^{2+}$ vs $SO_4$ + $HCO_3$ $^-$ (Fig. 6b) in the study area shows that most of the groundwater samples of the alluvial aquifer are clustered around and above the 1:1 line, indicating reverse ion exchange (El-Sayed et al. 2012).

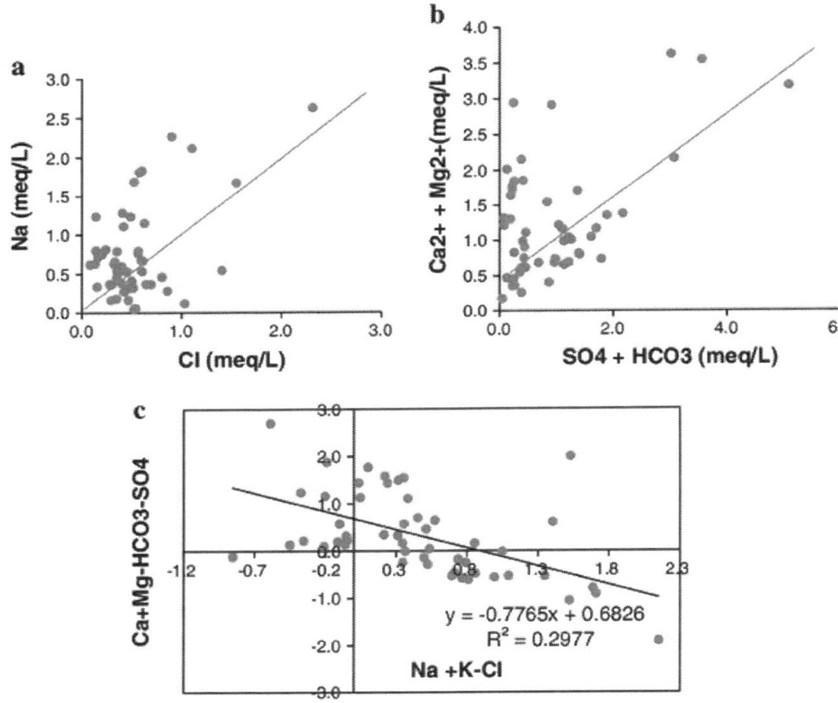


**Figure 6.** Graphs of different parameters: **a** Na vs. Cl, **b** Ca + Mg vs. $SO_4$ + $HCO_3$, **c** Ca + Mg–$HCO_3$–$SO_4$

The evidence for cation exchange is also given by bivariate plot of $Ca^{2+}$ + $Mg^{2+}$–$HCO_3$–$SO_4$ as a function of $Na^+$ + $K^+$–$Cl^-$, as shown in Fig. 6c. If cation exchange was not a significant reaction in the aquifer system, all data would plot close to the origin (McLean et al. 2000). But if cation exchange was an important composition controlling process, the relation between these two parameters should be linear with a slope of −1. Figure 6c indicates an increase in $Na^{+}$ + $K^{+}$ related to a decrease in $Ca^{2+}$ + $Mg^{2+}$ or an increase in $HCO_3$ + $SO_4$. All data plot close to a straight line with a slope of −0.776, indicating that some of the $Na^{+}$, $Ca^{2+}$and $Mg^{2+}$ participate in the ion-exchange reaction (Garcia et al. 2001). Furthermore, ion exchange and reverse ion exchange occurring in the groundwater were confirmed by two basic exchange indices, namely the chloro alkaline indices (CAI-1 and CAI-2):

$$\text{CAI-1} = [Cl^{-} - (Na^{+} + K^{+})]/Cl^{-} \quad (1)$$

$$\text{CAI-2} = [Cl^{-} - (Na^{+} + K^{+})]/(SO_4 + HCO_3 + CO_3 + NO_3) \quad (2)$$

(all values are expressed in meq/L). If there was ion exchange of $Na^{+}$ and $K^{+}$ from water with magnesium and calcium in the rock, then the exchange is known as direct when the indices are positive. If the exchange is in the reverse order, then the exchange is indirect and the indices are found to be negative (Schoeller 1967). The calculated chloro alkaline indices are presented in Table 1. It was observed that 75 % of the groundwater samples show negative ratios and thus exhibit a reverse ion exchange.

The area of study is characterised by shallow water table, underlain by very pyritic carbonaceous (peat, lignite and organic matter) wetland sediments. These sediments contain inorganic sulphides including pyrite and other iron-bearing minerals such as pyroxene, amphiboles, biotite, magnetite and olivine (Amadi et al.1987). Oxygen transported into such layers, either by convection transport or direct exposure to air by lowering of groundwater levels results in oxidation of pyrite. Oxidation of pyrite, dissolution of these iron-bearing minerals and the decomposition of organic matter result in the release of acid, iron etc. into the soil and groundwater system (Jeong 2001; Chirenje et al. 2007). In this study, the $SO_4/Cl$ ratio is used as an indicator to illustrate potential pyritic oxidation. $SO_4/Cl$ ratio ≥0.5 reflects pyrite oxidation. Table 1 shows that in areas with $SO_4/Cl$ ratios ≥0.5 also have high Fe concentrations in the groundwater due to oxidation of pyrite.

This results in a decrease in pH and thus explains in part the mild acidity and elevated concentration of iron observed in the groundwater system. Also, the low values of sulphate concentration in the groundwater (Table 1) are attributed to the removal of sulphate from the water by sulphate-reducing bacteria. The reduction of sulphate ions produces hydrogen sulphide; the unpleasant odour of which could easily be detected in the swamps of the Niger Delta (Amadi et al. 1987).

## 4.3 Dissolution and deposition

Mineral equilibrium calculations for groundwater are useful in predicting the presence of reactive minerals in the groundwater and estimating mineral reactivity (Deutsch 1997). Using the saturation index (SI) approach, it is possible to predict the reactive mineralogy of the subsurface from groundwater without collecting the samples of the solid phase and analysing the mineralogy (Deutsch 1997). In the present study, the calculated values of SI for calcite and dolomite of the groundwater samples range from −2.99 to 2.0 and −3.89 to 3.64 with averages of −0.32 and −0.27 (Fig. 7a, b), respectively. The groundwater samples were kinetically undersaturated to oversaturated with respect to calcite and dolomite, and were undersaturated with respect to gypsum (Fig. 7c).

## 4.4 Principal component analysis

The results of the principal component analyses are shown in Table 3. They include the loading for the rotated component matrix, the eigenvalues for each component, and the percent and cumulative percent of variance explained by each component and communality, indicating the proportion of variance of each variable controlled by the set of components. From Table 3, it can be seen that the first three principal components together account for 78.38 % of the total variance in the dataset, in which the first principal component explains 44.09 % of the total variance, the second principal component exhibits 23.34 % of the total variance, and the third principal component explains 10.95

% of the total variance. The concentrations of EC, TDS, TA, TH, $Na^+$, $K^+$, Ca and $HCO_3^-$ show high loadings (0.743–0.959), and the concentration of $Cl^-$has moderate positive loading (0.616) on the first principal component. On the second principal component, Fe and $NO_3^-$ concentrations have high positive loadings (0.867–0.903), and the concentration of $Mg^{2+}$shows moderate positive loading (0.576). The combinations of $Na^+$, $K^+$, Ca, $HCO_3^-$ and $Cl^-$ ions, influencing the high positive loadings observed on the first principal component (Table 3), suggest that the first principal component is associated with a combination of various hydrogeochemical processes that contribute to enrich more mineralized water (TDS).

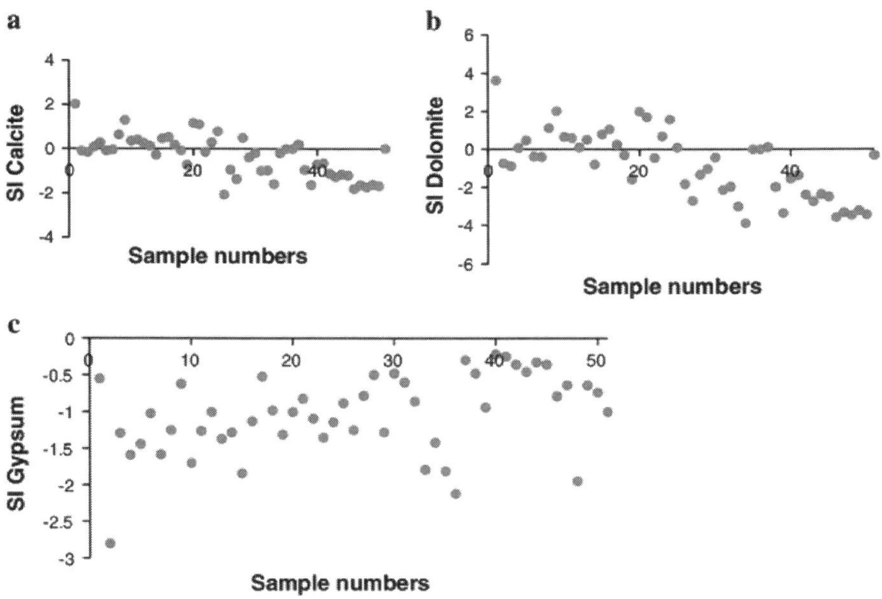


**Figure 7.** Saturation indices **a** Calcite, **b** Dolomite, **c** Gypsum

For instance, a high positive loading of $Na^+$ and $K^+$ ions indicates natural weathering of rock minerals and various ion-exchange processes in the groundwater system. The strong loading of $HCO_3^-$ ions with alkali and alkaline earth metals supports the view of natural weathering sources (Srivastava and Ramanathan2008). The high positive loading of $HCO_3^-$ is attributed to the reaction of feldspar minerals with carbonic acid in the presence of water (Elango et al. 2003). The concentrations of iron observed as high positive loadings on the second principal component (Table 3) are inferred to be related to the oxidation of inorganic sulphides including pyrite and leaching of iron-bearing minerals such as pyroxene, amphiboles, biotite, magnetite and olivine (Amadi et al. 1987). Nitrate has no significant lithologic source in the study area and it must be associated with the anthropogenic activities. Factor 3 explained that 10.95 % of the total variance is found to be associated with strong negative loading of $SO_4{}^{2-}$ and weak pH. This factor may be attributed to anthropogenic activities such as domestic waste discharge.

**Table 3.** Rotated component matrix of chemical data of groundwater samples

| Chemical variables | Principal components | | | Communality |
|---|---|---|---|---|
| | I | II | III | |
| pH | 0.351 | 0.297 | 0.401 | 0.372 |
| EC | 0.959 | 0.068 | 0.084 | 0.931 |
| TDS | 0.954 | 0.096 | 0.096 | 0.929 |
| TA | 0.791 | −0.260 | 0.249 | 0.755 |
| TH | 0.842 | 0.060 | 0.121 | 0.727 |
| $Na^+$ | 0.843 | −0.416 | −0.074 | 0.890 |
| $K^+$ | 0.829 | −0.280 | 0.002 | 0.765 |
| $Ca^{2+}$ | 0.842 | 0.250 | −0.302 | 0.863 |
| $Mg^{2+}$ | 0.489 | 0.576 | −0.325 | 0.676 |
| Fe | 0.070 | 0.903 | 0.003 | 0.820 |
| $Cl^-$ | 0.616 | 0.112 | −0.609 | 0.764 |
| $SO_4{}^{2-}$ | −0.019 | −0.178 | −0.851 | 0.756 |
| $NO^{3-}$ | 0.177 | 0.867 | 0.117 | 0.796 |
| $HCO_3{}^-$ | 0.743 | −0.531 | 0.254 | 0.899 |
| Eigenvalues | 6.613 | 3.502 | 1.642 | |
| % of variance explained by component | 44.085 | 23.344 | 10.946 | |
| Cumulative % of variance | 44.085 | 67.429 | 78.375 | |

# 5. CONCLUSION

The present study demonstrates the usefulness of factor analysis in combination with conventional graphical techniques in interpreting hydrogeochemical data to evaluate the major factors influencing groundwater composition. The degree to which the groundwater composition was influenced by various factors had been assessed by examining the factor scores. The hydrogeochemical characteristics of the groundwater revealed that the groundwater is slightly acidic and major anions and cations were below WHO prescribed limits for drinking water except iron. Prevalent water types in the study are: Ca–Cl, Ca–$HCO_3$–Cl, Ca–Na–Cl, Ca–Na–$HCO_3$–Cl, Ca–Na–$HCO_3$ and Na–$HCO_3$. The results of the multivariate statistical analysis in combination with conventional graphical methods indicate that the chemical composition of groundwater of the study area is strongly influenced by rock–water interaction, oxidation of inorganic sulphides including pyrite, weathering of silicates group of minerals enhanced by reverse ion-exchange processes. Weathering of silicate minerals controls the major ion chemistry of sodium, potassium, calcium and magnesium. The enhanced concentration of iron in the groundwater in the area is due to oxidation of pyrite and leaching of iron-bearing minerals. Nitrate and chloride have no significant lithologic source in the study area and must be associated with the anthropogenic activities such as sewage wastes.

# ACKNOWLEDGMENTS

The authors are grateful to all home owners who allowed us access to their premises to collect water samples from their domestic boreholes. Special thanks to Mr. Suoye Diete-spiff of the Central Research Laboratory, Niger Delta University for the physico-chemical analysis and two anonymous reviewers whose comments improved the quality of this paper.

# REFERENCES

1. Akpokodje EG (1986) A method of reducing the cement content of two stabilized Niger Delta soils. Q J Eng Geol London 19:359–363

2. Amadi PA, Ofoegbu CO, Morrison T (1987) Hydrogeochemical assessment of groundwater quality in parts of the Niger Delta, Nigeria. Environ Geol Water Sci 14:195–202

3. Amajor LC (1991) Aquifers in the benin formation (Miocene-Recent), Eastern Niger Delta, Nigeria. lithostratigraphy, hydraulics and water quality. Environ Geol Water Sci 17(2):85–101

4. APHA (1995) Standard methods for the examination of water and waste water (17th ed) American Public Health Association, Washington

5. Ashley RP, Lloyd JW (1978) An example of the use of factor analysis and cluster analysis in groundwater chemistry interpretation. J Hydrogeol 39:355–364

6. Cerling TE, Pederson BL, Damm KLV (1989) Sodium–Calcium ion exchange in the weathering of shales: implications for global weathering budgets. Geology 17:552–554

7. Chirenje T, Epstein C, Mueller R (2007) Water quality issues in the outer coastal plains New Jersy. In: Sarkar et al. (eds) Developments in Environmental Science, 5, pp 561–589

8. Davis JC (1986) Statistics and data analysis in geology, 2nd edn. Willey, New York

9. Davis SN, De Wiest RJM (1966) Hydrogeology. Wiley, New York

10. Deutsch WJ (1997) Groundwater geochemistry: fundamentals and application to contamination. CRC, Boca Raton

11. Durotoye B (1989) Quaternary sediments in Nigeria: In Kogbe (ed.) Geology of Nigeria, Rock View (Nigeria) Ltd. Publishers, Jos

12. Elango L, Rannan R, Senthil KM (2003) Major ion chemistry and identification of hydrogeochemical processes of groundwater in a part of Kancheepuram district, Tamil Nadu, India. J Environ Geosci 10:157–166

13. El-Sayed MH, Moustafa M, El-Fadl MA, Shawky H (2012) Impact of hydrochemical processes on groundwater quality, Wadi Feiran, South Sinai, Egypt. Australian J Basic Appl Sci 6:638–654

14. Fisher RS, Mulican WF (1997) Hydrochemical evolution of sodium-sulfate and sodium-chloride groundwater beneath the Northern Chihuahuan desert, Trans Pecos, Rexas, USA. Hydrogeol J 10:455–474

15. Garcia MG, del Hildalgo VM, Blessa MA (2001) Geochemistry of groundwater in the alluvial plain of Tucuman province, Argentina. Hydrogeol J 9:597–610

16. Gibbs RJ (1970) Mechanisms controlling World's water chemistry. Science 170:1088–1090

17. Guler C, Thyne GD, McCray JE, Turner AK (2002) Evaluation of graphical and multivariate statistical method for classification of water chemistry data. Hydrogeol J 10(4):455–474

18. Hem JD (1989) Study and interpretation of the chemical characteristics of natural water, 3rd edition. US Geol Surv Water Supply Paper 2254:263p

19. Jeong CH (2001) Mineral–water interaction and hydrogeochemistry in Samkwang mine area, Korea. Geochem J 35:1–12

20. Kogbe CA (1989) The cretaceous and paleogene sediments of southern Nigeria. In: Kogbe (ed.) Geology of Nigeria, Elizabethan Press, Lagos, pp 311–334

21. Krishnaraj S, Murugesan V, Vijayaraghaavan K, Sabarathinam C, Paluchamy A, Ramachandran M (2011) Use of hydrochemistry and stable isotopes as tools for groundwater evolution and contamination investigations. Geosciences 1:16–25

22. Lawrence FW, Upchurch SB (1982) Identification of recharge areas using geochemical factor analysis. Groundwater 20:680–687

23. Lee JY, Cheon JY, Lee KK, Lee SY, Lee MH (2001) Statistical evaluation of geochemical parameter distribution in a groundwater system contaminated with petroleum hydrocarbons. J Environ Qual 30:1548–1563

24. Lloyd JW, Heathcote JA (1985) Natural inorganic hydrochemistry in relation to groundwater, an introduction. Clarendon Press, Oxford

25. McLean W, Jankowski J, Levitt N (2000) Groundwater quality and sustainability in an alluvial aquifer, Australia. In: Sililo O (ed) Groundwater, past achievements and future chanllenges. Rotterdam, Balkenna, pp 567–573

26. Ofoegbu CO (1989) A review of the geology of the Benue Trough, Nigeria. J Afr Earth Sci 2:283–291

27. Okiongbo KS, Douglas RK (2013) Hydrogeochemical analysis and evaluation of groundwater quality in Yenagoa City and environs, Southern Nigeria. Ife J Sci 15:209–222

28. Okiongbo KS, Ogobiri G (2011) Geoelectric investigation of groundwater resources in parts of Bayelsa State, Nigeria. Res J Environ Earth Sci 3(6):620–624

29. Onyeagocha AC (1980) The petrology and depositional environment of the Benin Formation. Nigerian J Mining Geol 17:147–151

30. Reeder SW, Hitchon B, Levinson AA (1972) Hydrogeochemistry of the surface waters of the Mackenzie River drainage basin, Canada I. Factors controlling inorganic composition. Geochim Cosmochim Acta 36:825–865

31. Rogers RJ (1989) Geochemical comparison of groundwater in areas of New England, New York, and Pennsylvania. Groundwater 27:690–712

32. Ruiz F, Gomis V, Blasco P (1990) Application of factor analysis to the study of a coastal aquifer. Hydrogeol J 119:169–177

33. Schoeller H (1967) Qualitative evaluation of groundwater resources. In: Methods and techniques of groundwater investigation and development. Water Research, Series-*33:* UNESCO, pp 45–52

34. Short KC, Stauble AJ (1967) Outline of the geology of the Niger Delta. Bull AAPG 51:761–779

35. Srinivasamoorthy K, Chidambaram S, Prasanna MV, Vasanthavihar M, Peter J, Anandhan P (2008) Identification of major sources controlling groundwater chemistry from hard rock terrain—a case study from Mettur taluk, Salem district, Tamil Nadu, India. J Earth Syst Sci 117:49–58

36. Srivastava SK, Ramanathan AL (2008) Geochemical assessment of groundwater quality in vicinity of Bhalswa landfill, Delhi, India, using graphical and multivariate statistical methods. Environ Geol 53:1509–1528

37. World Health Organisation (WHO) (2004) Guidelines for drinking water quality: incorporating 1st and 2nd vol. 1 recommendations 3rd ed., Geneva

38. Wright JB, Hasting DA, Jones WB, Williams HR (1985) Geology and mineral resources of West Africa, Allen and Unwin Limited, UK, pp 107

10

# HYDRO-CHEMICAL EVOLUTION OF GROUNDWATER AND MIXING BETWEEN AQUIFERS: A STATISTICAL APPROACH BASED ON MAJOR IONS

**Linhua Sun, Herong Gui**

School of Earth Science and Engineering, Suzhou University, Suzhou 234000, Anhui, China

## ABSTRACT

Geochemical analysis is a useful tool in hydrogeological assessment, particularly in constructing a conceptual model of a hydrogeological system. In this study, major ion concentrations of 53 groundwater samples from the coal-bearing aquifer in the Qidong coal mine, northern Anhui Province of China have been processed by statistical analysis for understanding either hydro-chemical characteristics or hydrological evolution, which will be useful for the safety of coal mining. The results suggest that most of the samples are Na–$SO_4$ and Na–$HCO_3$ types, and their hydro-chemical compositions are mainly controlled by dissolution of more soluble minerals (e.g. calcite) and weathering of silicate minerals (e.g. plagioclase). Two groups of samples have been subdivided by quantile and scatter plots of factor scores, one is related to different degrees of water–rock interactions and another is related to groundwater mixing. Moreover, four end members have been identified and the mixing calculation suggests that the groundwater samples affected by mixing have 20–100 % contribution from the loose layer aquifer (LA), and therefore, groundwater from the LA in the coal mine should be taken seriously during coal mining. The study demonstrated that statistical analysis is useful for connecting the hydrochemistry of groundwater with hydrological evolution of the aquifer.

## KEYWORDS

Groundwater, Hydrochemistry, Statistical analysis, Water–rock interaction, Mixing, Coal mine

## 1. INTRODUCTION

Understanding the mixing between groundwater and other water bodies is an important work for groundwater management, because it is a common phenomenon and important process governing groundwater chemistry in many natural systems, and a large number of studies have been processed (Schramke et al. 1996; Johannesson et al. 1997; Laaksoharju et al. 1999, 2008; Rueedi et al. 2005; Long and Valder 2011; Valder et al. 2012; Morales-Casique 2012). Among these studies, geochemistry is considered to be a useful tool in hydrogeological assessment, particularly in constructing a conceptual model of a hydrogeological system. Previous studies revealed that the chemical compositions of groundwater are controlled by complex processes (e.g. recharge, reaction and mixing). However, in some situations, mixing can explain the leading order variability in geochemical composition, while chemical reactions explain the remaining variability (Laaksoharju et al. 1999, 2008; Morales-Casique 2012).

Mixing calculations based on hydrochemistry have been successfully applied in several hydrogeological settings for better understanding of the hydrological evolution. Taking for instance, mixing of three end members related to shallow and deep groundwater systems have been postulated by Ramos-Leal et al. (2007) for explaining the geochemical variations in an aquifer of Central Mexico, whereas mixing has been applied for understanding the hydro-chemical evolutions in Sweden and Canada in combination with water–rock interactions (Laaksoharju et al. 2008). A detailed review of other applications of mixing calculations can be found in Nakaya et al. (2007). These studies demonstrated that the concentration of solutes of mixtures between two end members follows a linear relation, while chemical reactions of reactive constituents introduce significant non-linear effects (Wigley and Plummer 1976).

Hydrogeological studies of groundwater are important work in coal mines, because they can provide information to understand the hydrological evolution of the area, which is essential for water disaster controlling during coal mining (e.g. water source identification after water inrush). Therefore, a large number of hydro-chemical studies including major ions, rare earth elements and stable isotopes have long been carried out in northern Anhui Province of China, as the area is an important energy base with huge coal production every year (more than 30 million tons) (e.g. Gui and Chen 2007, Gui et al. 2011; Sun et al. 2011, Sun and Gui 2012). However, the groundwater mixing between aquifers has not been well-understood, which is important for water inrush controlling in coal mines (e.g. multi-water source identification) (Sun2013).

In this study, we report a case study in the Qidong coal mine, northern Anhui Province, China. The major ion concentrations of groundwater samples from the coal-bearing aquifer (CA) have been processed by statistical analysis and mixing model. The goals of this study include: (1) understanding the hydro-chemical characteristics of the groundwater and (2) identifying the end members affecting the chemical compositions of groundwater. The study can provide useful information for understanding the hydrological evolution of the aquifer. Moreover, it is important for future water disaster controlling in the coal mine because it can reveal the potential hydraulic connection between aquifers.

# 2. SITE DESCRIPTION

The Qidong coal mine is located 20 km south to Suzhou City, northern Anhui Province of China and belongs to Suxian coal field (Fig. 1). The width of the coal mine is 3–8 km from south to north, and the length is 10–13 km from east to west, the total area is 47.3 km$^2$. The climate in the area is warm and semi-humid. The annual average temperature is 14.4 °C and the average annual rainfall is 834 mm, most of which concentrates between July and August.

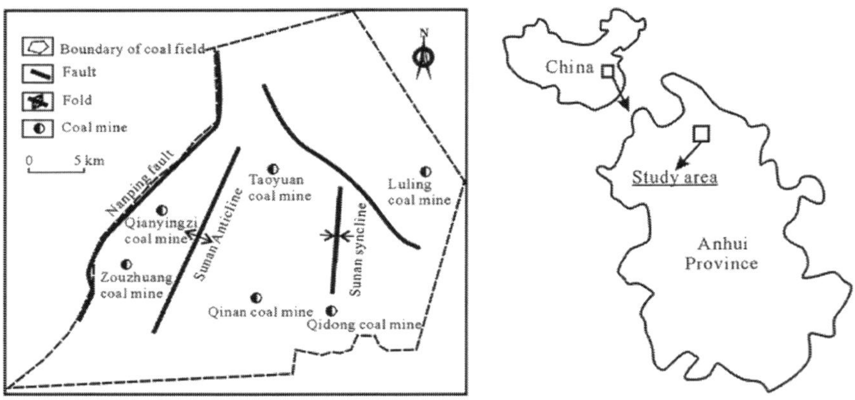


**Figure 1.** Location of the Qidong coal mine

Previous explorations revealed that the groundwater system in the mine can be divided into three major aquifer systems from shallow to deep: loose layer aquifer system (LA), coal-bearing sandstone aquifer system (CA) and the underlying limestone aquifer system (TA). Each major aquifer system can be subdivided into certain small secondary aquifers and their characteristics are as follows:

LA: containing four secondary aquifers. The first aquifer is buried in shallow environment (<40 m) and the wall rocks are dominated by siltstone and clay. The second aquifer is located between 78 and 101 m depth and the main rock types in the aquifer are siltstone and fine sandstone. The third aquifer is buried between the depth of 185 and

200 m and the host rocks are mainly composed of siltstone and median sandstone. The depth of the fourth aquifer is 285–400 m and the wall rocks are mainly composed of conglomerate, sandstone and siltstone. Quartz sandstone, limestone, flint are main constitutes in the conglomerates. The fourth aquifer is considered to be a potential threat for coal mining, as it has been proven to be the water supplier for the water inrush in 2011 (Gui and Chen 2007).

CA: three secondary aquifers have been subdivided. The first one is named as K3 aquifer, which is mainly composed of gray–white median and fine sandstones with a thickness of 12 m, and the coals related to it include 1st, 2nd and 3rd coal seams. The second aquifer is related to 7th–9th coal seams, the main rock types are median and fine sandstones. The third one is a 10th coal seam related aquifer, which is composed of fine sandstones with siltstone and clay. In summary, CA is a close aquifer system and has limited water storage.

TA: two secondary limestone aquifers have been identified in the aquifer system, including the upper Taiyuan TA and lower Ordovician TA. The total thickness of the Taiyuan Formation is 193 m and there are 14 layers of limestone with a total thickness of 83 m. The water is stored in the dissolution-related caves. During the mining of the 10th coal seam, the water from TA is considered to be a threat to coal mining safety, due to the close distance between 10th coal seam and TA.

# 3. MATERIALS AND METHODS

All of the samples were collected from the tunnel of coal mining in the Qidong coal mine during Jan 2011 and Dec 2011. A total of 53 samples have been collected and they were related to the 3rd, 4th and 6th–9th coal seams. Concentrations of seven kinds of major ions ($Na^+$, $Ca^{2+}$, $Mg^{2+}$, $Cl^-$, $SO_4{}^{2-}$, $HCO_3{}^-$ and $CO_3{}^{2-}$), pH values and total dissolved solids (TDS) have been analyzed, and the analytical methods are as follows: $Na^+$, $Ca^{2+}$, $Mg^{2+}$, $Cl^-$ and $SO_4{}^{2-}$ were analyzed by ion chromatography, whereas $HCO_3{}^-$ and $CO_3{}^{2-}$ were analyzed by acid–base titration in the

Engineering and Technological Research Center of Coal Exploration, Anhui Province, China.

Major ion concentrations were first processed by Aquachem (version 3.7) for classification of hydro-chemical facies. Factor analysis with maximum likelihood extraction in SPSS (version 16) was performed for understanding the source of major ions in combination with quantile plots (*Q–Q* plots) of major ion concentrations. Moreover, *Q–Q* and scatter plots of factor scores were used for sample grouping, and the end members responsible for the hydro-chemical compositions of the groundwater samples were identified by scatter plots of factor scores with extreme values in combination with hydrological conditions of the coal mine, as well as correlation between major ions, similar to the principle component analysis as reported by Laaksoharju et al. (2008). Moreover, a linear calculation of factor scores of end members was used for quantifying the mixing status of the groundwater samples.

# 4. RESULTS AND DISCUSSIONS

## 4.1 Hydro-chemical compositions

The statistics of hydro-chemical compositions of groundwater samples in this study are listed in Table 1. As can be seen from the table, most of the groundwater samples are dominated by $Na^+$ over other cations, whereas $HCO_3^-$ and $SO_4^{2-}$ are the dominant anions. Moreover, all of the major ions have high coefficients of variations (Table 1), indicating that the hydro-chemical compositions of the samples show a high variability among the study area. In comparison with previous studies in the study area (e.g. Sun and Gui2012), the groundwater samples in this study have lower $Na^+$, $Cl^-$ and $HCO_3^-$, but higher $Ca^{2+}$, $Mg^{2+}$ and $SO_4^{2-}$ than the groundwater samples from the similar CA in the Renlou coal mine, indicating different hydrological conditions (rocks and water rock interactions).

**Table 1.** Statistics of hydro-chemical compositions (mg/L) of groundwater samples

| | pH | $Na^+$ | $Ca^{2+}$ | $Mg^{2+}$ | $Cl^-$ | $SO_4^{2-}$ | $HCO_3^-$ | $CO_3^{2-}$ | TDS |
|---|---|---|---|---|---|---|---|---|---|
| *N* | 53 | 53 | 53 | 53 | 53 | 53 | 53 | 53 | 53 |
| Min | 5.6 | 204 | 0 | 4 | 113 | 2 | 255 | 0 | 735 |
| Max | 9.7 | 940 | 203 | 106 | 256 | 1,506 | 1,358 | 409 | 2,781 |
| Mean | 8.2 | 380 | 69 | 45 | 202 | 380 | 581 | 37 | 1,404 |
| SD | 0.68 | 175 | 71 | 39 | 49 | 348 | 258 | 72 | 434 |
| CV | 0.08 | 0.46 | 1.04 | 0.86 | 0.24 | 0.92 | 0.44 | 1.93 | 0.31 |
| *p* value test | <0.01 | <0.01 | <0.01 | <0.01 | <0.01 | <0.01 | <0.01 | <0.01 | <0.01 |
| RL | | 731 | 22.3 | 17.9 | 566 | 72.2 | 915 | 41.0 | |

*SD* and *CV* mean standard deviation and coefficient of variation, respectively. *p* value obtains from Anderson–Darling test. *RL* is the mean concentration of groundwater samples from the CA in Renlou coal mine (after Sun and Gui 2012)

Based on the classification of the Piper and Durov diagrams (Fig. 2), all of the groundwater samples in this study can be subdivided into four kinds of hydro-chemical facies: including Ca–$SO_4$ (4), Na–Cl (3), Na–$SO_4$(21) and Na–$HCO_3$ (25). It can also be obtained from Fig. 2 that the groundwater samples in this study can be classified into two groups with different chemical signatures: the first one (G1) is dominated by Na–$HCO_3$and Na-Cl types, which is considered to be representative of groundwater in the CA, whereas group two with higher $Ca^{2+}$, $Mg^{2+}$ and $SO_4{}^{2-}$ concentrations is considered to be affected by other aquifers (see below). Moreover, Fig. 2 also gives information that their chemical characteristics are affected by three end members with $HCO_3{}^{-}$, $SO_4{}^{2-}$ and $Cl^{-}$ enrichment, respectively.

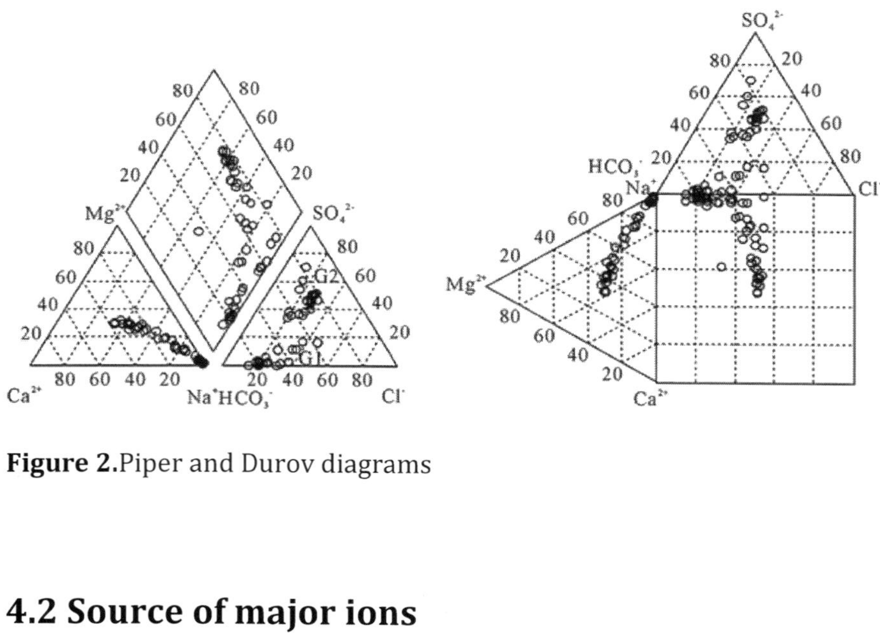


**Figure 2.**Piper and Durov diagrams

## 4.2 Source of major ions

Previous studies revealed that probability graphs of element concentrations are good indicators of sources (Reimann et al. 2005). In this study, more than one inflection points have been identified for all of the major ion concentrations (Fig. 3), implying that all of these major ions have more than one source. Take an instance, two sources are considered to be responsible for the variation of $Na^{+}$ concentrations because one inflection point can be identified from the *Q*–*Q* plot of $Na^{+}$ concentrations. This conclusion is further supported by their *p* values of

Anderson–Darling test that no major ion has $p$ values higher than 0.05 (under 95 % confidence, Table 1).

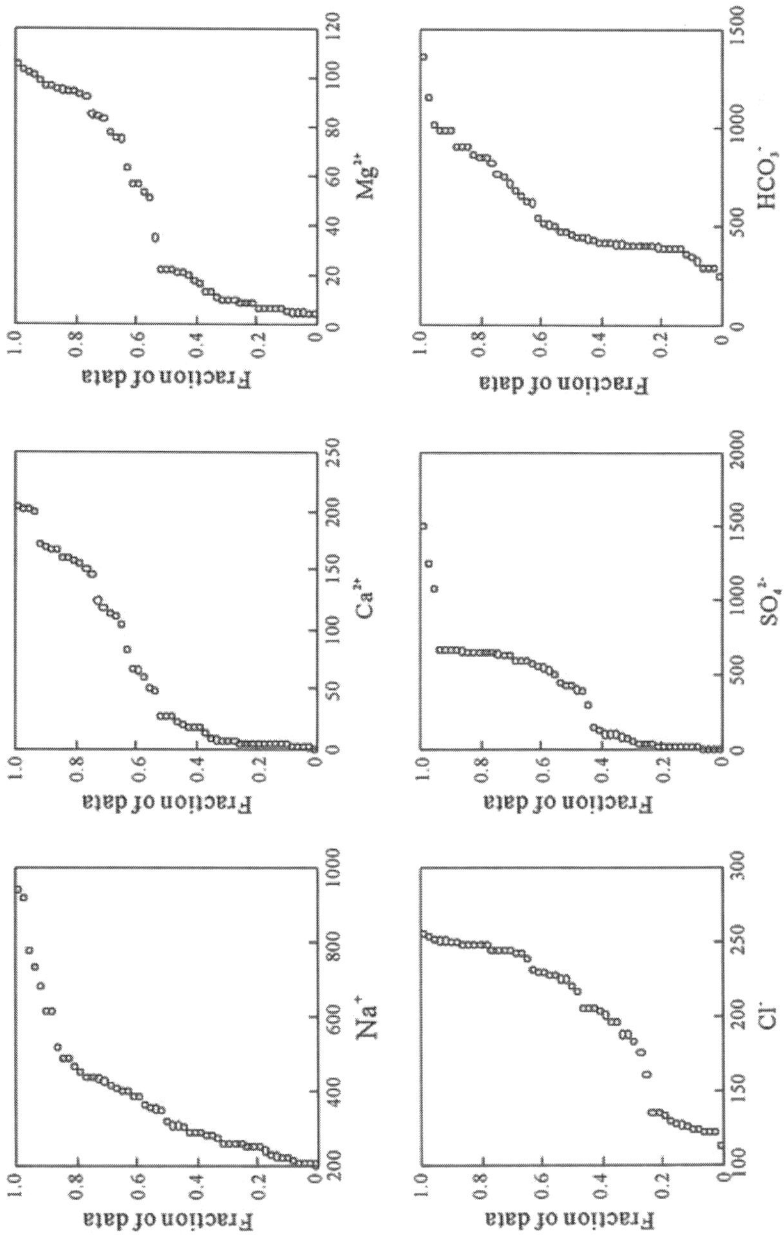


**Figure 3.** $Q$–$Q$ plots of major ion concentrations

Factor analysis is a commonly used statistical method. It has been used for classification, simplification of the data and finding the most important variables in the data set. During geochemical studies, factor analysis has long been used for tracing elemental sources (e.g. Cloutier et al. 2008; Zhou et al. 2008; Yalcin et al. 2010). In this study, factor analysis is processed using the major ion concentrations ($Na^+$, $Ca^{2+}$, $Mg^{2+}$, $Cl^-$, $SO_4^{2-}$ and $HCO_3^-$), and the result is shown in Fig. 4. Two factors with initial eigen value higher than one have been obtained.

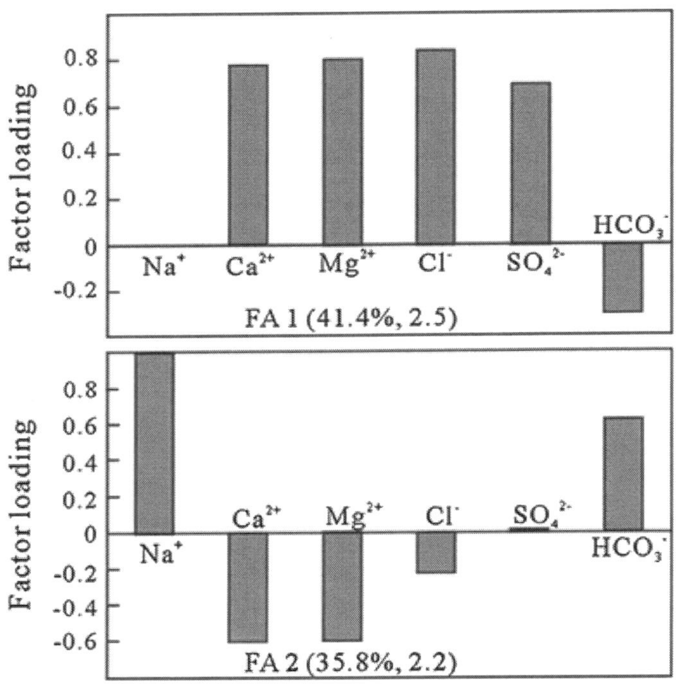


**Figure 4.** Bar charts of factor loadings

After varimax rotation, the total variance explanation is 77.2 %. The first factor (FA 1) accounts for 41.4 % of the total explanation and is dominated by $Ca^{2+}$, $Mg^{2+}$, $Cl^-$ and $SO_4^{2-}$, whereas the second factor (FA 2) accounts for 35.8 % information and is dominated by $Na^+$ and $HCO_3^-$. This result is similar to the results obtained from the Renlou (Sun and

Gui 2012) and the Wolonghu coal mine (Sun 2013). In combination with the wall rock compositions in the study area, as well as the understanding of previous studies, the FA 1 is considered to be representative of dissolution of more soluble minerals, such as calcite, dolomite and chloride, whereas the FA 2 is considered to be weathering of silicate mineral because only weathering of silicate minerals can generate $Na^+$ (e.g. plagioclase) and $HCO_3^-$ simultaneously (Meybeck 1987).

## 4.3 End member identification and mixing calculation

As mentioned in the "Site description", three sources are considered to be potential sources that can affect the chemical compositions of groundwater in the CA, including LA, TA and itself. However, these aquifers have different wall rock compositions, and therefore, the groundwater from them can be identified through their chemical compositions, based on the fact that they are predominantly controlled by water–rock interactions under the condition without groundwater mixing (Sun and Gui 2011, Sun and Gui 2012; Gui et al. 2011).

As can be seen from Fig. 5, the groundwater samples have $Ca^{2+}/Na^+$, $HCO_3^-/Na^+$ and $Mg^{2+}/Na^+$ ratios that range between the fields of evaporate dissolution and silicate weathering, and no sample shows contribution from carbonate dissolution. Therefore, contribution from TA is not expected, and the main contributors for the hydrochemistry of the groundwater in this study are considered to be LA and CA.

Factor score plots can provide important information about groundwater mixing and end member identification (Laaksoharju et al. 1999, 2008). Although they use principle component scores for solving this issue, here we use factor scores instead. It can be found in *Q–Q* plots of factor scores (Fig. 6) that there are two inflection points in factor score 1 and 2. And, according to these inflection points, the samples in scatter plots of factor scores (Fig. 7a) can be subdivided into three groups and three trends can be identified:

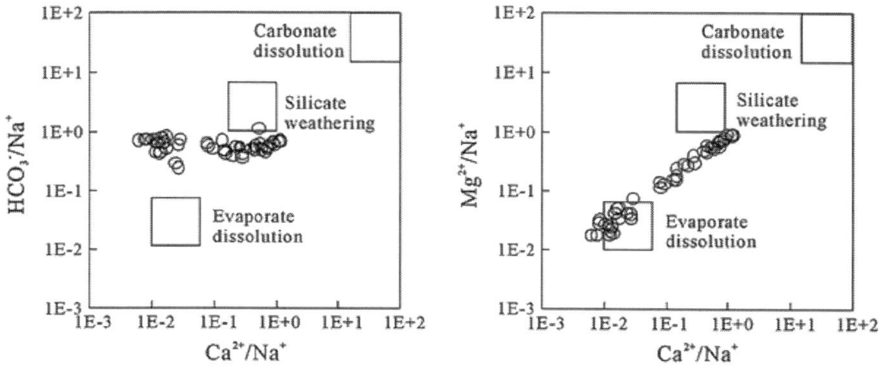


**Figure 5.**Na⁺ normalized $Ca^{2+}$–$HCO_3^-$ and $Ca^{2+}$–$Mg^{2+}$ plots

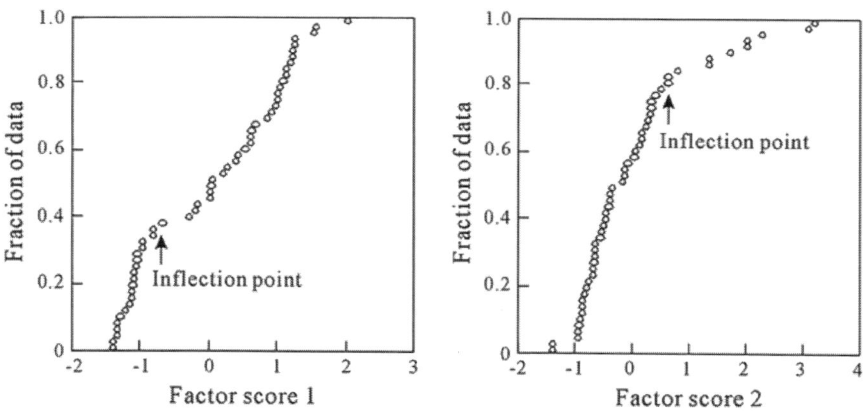


**Figure 6.***Q*–*Q* plots of factor scores

G1, samples located between Q14 and Q11, which indicates lower degree of water–rock interactions (depletion of contributions from more soluble and silicate minerals); G2, samples located between Q14 and Q50, which indicates higher degree of water–rock interactions (enrichment of contributions from more soluble and silicate minerals); G3, samples located between Q14 and Q29, which indicates more contribution from more soluble minerals but less contribution from silicate minerals, and cannot be explained by different degrees of water–rock interactions. Moreover, because the contribution from TA is not important for the chemical compositions of the groundwater samples in this study, the trend from Q14 to Q29 is considered to be the express of groundwater mixing between LA and CA, because more carbonates can only be found in the wall rocks of the fourth aquifer in LA except for TA.

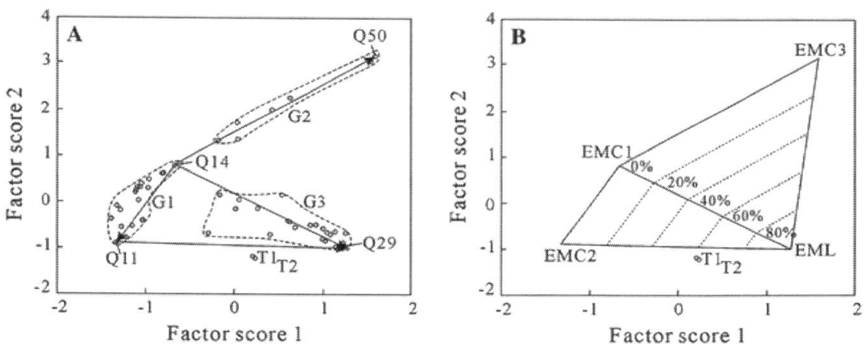


**Figure 7.a** Plots of factor scores and **b** mixing calculation using end members defined by factor analysis

To confirm this consideration, major ion concentrations of two samples (T1 and 2) reported by Gui and Chen (2007) during water inrush in 2001 have been analyzed simultaneously with the data in this study. It should be noticed that these two samples had been identified as mixing water supplied by LA and CA, because during the water inrush, the water table of LA is decreasing fast. As can be seen from Fig. 7a, these two samples are plotted near the field of G3 and located between sample Q11 and Q29, which is similar to the previous understanding that they are mixing water of LA and CA.

Therefore, the hydrochemistry of groundwater samples in this study can be explained by two kinds of hydrological mechanisms: different degrees of water–rock interactions, which is expressed by Q11–Q14–Q50 trend, another one is groundwater mixing between LA and CA, which is expressed by Q14/Q11–Q29 trend. Moreover, because the chemical compositions of the groundwater from CA have a broad range, samples including Q14, Q11 and Q50 are selected as end members of CA (EMC1, 2 and 3) and Q29 is selected as end member of LA (EML) (Fig. 7b).

It can be seen in Fig. 7a that G3 group samples are mixtures of Q11, Q14 and Q29. Therefore, the mixing calculations among the three samples were processed by two kinds of methods: one is mixing using Aquachem software and another is linear calculations of factor scores of end members. The results are listed in Table 2 and shown in Fig. 7b. As can be seen from the table and figure, these two kinds of methods give different results: the linear calculations suggest that the samples in 2011 (T1 and T2) have ~60 % contributions from the Q29 (EML), whereas the mixing calculations by Aquachem software suggest 44 and 42 % contributions from the EML. These differences suggest that not only the mixing, but also other processes (e.g. chemical reaction) must have been undertaken.

**Table 2.** Results of mixing calculations using Aquachem software

| End member | $Na^+$ | $Ca^{2+}$ | $Mg^{2+}$ | $Cl^-$ | $SO_4^{2-}$ | $HCO_3^-$ |
|---|---|---|---|---|---|---|
| Q11 (EMC2) | 224 | 28 | 22 | 205 | 103 | 290 |
| Q29 (EML) | 206 | 201 | 102 | 249 | 663 | 399 |
| T1 (original) | 129 | 180 | 70 | 299 | 282 | 448 |
| T1 (58 %) | 216 | 101 | 56 | 223 | 338 | 336 |
| T2 (original) | 128 | 179 | 72 | 302 | 289 | 452 |
| T2 (56 %) | 216 | 104 | 57 | 224 | 349 | 338 |

T1 (58 %) and T2 (56 %) are the results of mixing calculations by Aquachem software. 58 and 56 % are the proportions of EMC2 obtained by mixing calculation

# 5. CONCLUSIONS

Major ion concentrations of groundwater samples from the CA in the Qidong coal mine, northern Anhui Province of China have been processed by statistical analysis for the understanding of their hydro-chemical characteristics and hydrological evolution, and the following conclusions have been made:

The groundwater samples are dominated by Na–$SO_4$ and Na–$HCO_3$ types, to a lesser extent, Ca–$SO_4$ and Na–Cl types. *Q–Q* plots of major ion concentrations suggest that they are originated from multi-sources, which is further supported by factor analysis that dissolution of more soluble minerals and weathering of silicate minerals are main contributors.

Plots of sodium-normalized calcium, magnesium and bicarbonate indicate that TA is not important for the hydrochemistry of the groundwater, whereas *Q–Q* plots and scatter plots of factor scores imply that most of the samples are controlled by different degrees of water–rock interactions, whereas others have been affected by groundwater mixing, and mixing calculation of end members suggests that these samples have 20–100 % contribution from LA.

The study demonstrated that statistical analysis of hydrochemistry can be used for understanding of the hydro-chemical and hydrological evolution of the groundwater system (e.g. water–rock interaction and groundwater mixing) and, during coal mining in the Qidong coal mine, the threats posed by LA should be taken seriously.


# ACKNOWLEDGMENTS

This work was financially supported by National Natural Science Foundation of China (41302274 and 41373095) and the Natural Science Foundation of Anhui Province (1308085QE77).

## REFERENCES

1. Cloutier V, Lefebvre R, Therrien R, Savard MM (2008) Multivariate statistical analysis of geochemical data as indicative of the hydrogeochemical evolution of groundwater in a sedimentary rock aquifer system. J Hydrol 353(3):294–313

2. Gui HR, Chen LW (2007) Hydrogeochemistric evolution and discrimination of groundwater in mining district. Geological Publishing House, Beijing

3. Gui HR, Sun LH, Chen LW, Chen S (2011) Rare earth element geochemistry of groundwater from a deep seated sandstone aquifer, northern Anhui province, China. Min Sci Technol 21(4):477–482

4. Johannesson KH, Stetzenbach KJ, Hodge VF (1997) Rare earth elements as geochemical tracers of regional groundwater mixing. Geochim Cosmochim Acta 61(17):3605–3618

5. Laaksoharju M, Skårman C, Skårman E (1999) Multivariate mixing and mass balance (M3) calculations, a new tool for decoding hydrogeochemical information. Appl Geochem 14(7): 861–871

6. Laaksoharju M, Gascoyne M, Gurban I (2008) Understanding groundwater chemistry using mixing models. Appl Geochem 23(7):1921–1940

7. Long AJ, Valder JF (2011) Multivariate analyses with end-member mixing to characterize groundwater flow: wind cave and associated aquifers. J Hydrol 409(1):315–327

8. Meybeck M (1987) Global chemical weathering of surficial rocks estimated from river dissolved loads. Am J Sci 287(5):401–428

9. Morales-Casique E (2012) Mixing of groundwaters with uncertain end-members: case study in the Tepalcingo-Axochiapan aquifer, Mexico. Hydrogeol J 20(3):605–613

10. Nakaya S, Uesugi K, Motodate Y, Ohmiya I, Komiya H, Masuda H, Kusakabe M (2007) Spatial separation of groundwater flow paths from a multi-flow system by a simple mixing model using stable isotopes of oxygen and hydrogen as natural tracers. Water Resour Res 43(9):W09404.

11. Ramos-Leal JA, Martinez-Ruiz VJ, Rangel-Mendez JR, de la Torre MA (2007) Hydrogeological and mixing process of waters in aquifers in arid regions: a case study in San Luis Potosi Valley, Mexico. Environ Geol 53(2):325–337

12. Reimann C, Filzmoser P, Garrett RG (2005) Background and threshold: critical comparison of methods of determination. Sci Total Environ 346(1):1–16

13. Rueedi J, Purtschert R, Beyerle U, Alberich C, Kipfer R (2005) Estimating groundwater mixing ratios and their uncertainties using a statistical multi parameter approach. J Hydrol 305(1):1–14

14. Schramke JA, Murphy EM, Wood BD (1996) The use of geochemical mass-balance and mixing models to determine groundwater sources. Appl Geochem 11(4):523–539

15. Sun LH, Gui HR (2012) Establishment of water source discrimination model in coal mine by using hydrogeochemistry and statistical analysis: a case study from Renlou Coal Mine in northern Anhui Province, China. J Coal Sci Eng 18(4):385–389

16. Sun LH (2013) Statistical analysis of hydrochemistry of groundwater and its implications for water source identification: a case study. Arab J Geosci.

17. Sun LH, Gui HR, Chen S (2011) Rare earth element geochemistry of groundwaters from coal bearing aquifer in Renlou coal mine, northern Anhui Province, China. J Rare Earths 29(2):185–192

18. Valder JF, Long AJ, Davis AD, Kenner SJ (2012) Multivariate statistical approach to estimate mixing proportions for unknown end members. J Hydrol 460–461:65–76

19. Wigley TML, Plummer LN (1976) Mixing of carbonate waters. Geochim Cosmochim Acta 40(9):989–995

20. Yalcin MG, Tumuklu A, Sonmez M, Erdag DS (2010) Application of multivariate statistical approach to identify heavy metal sources in bottom soil of the Seyhan River (Adana), Turkey. Environ Monit Assess 164(1–4):311–322

21. Zhou J, Ma D, Pan J, Nie W, Wu K (2008) Application of multivariate statistical approach to identify heavy metal sources in sediment and waters: a case study in Yangzhong, China. Environ Geol 54(2):373–380

# INDEX

## T



## W



## Y